Communication Acoustics an Introduction to Speech, Audio and Psychoacoustics

Communication Acoustics an Introduction to Speech, Audio and Psychoacoustics

Editor

Abramo Adessi

Communication Acoustics An Introduction to Speech, Audio and Psychoacoustics

Edited by **Abramo Adessi**

Printed in 2017

ISBN: 978-1-68117-112-8
Library of Congress Control Number: 2015954091

© 2016 by
SCITUS Academics LLC,
616, Corporate Way, Suite 2, 4766,
Valley Cottage, NY 10989

www.scitusacademics.com

Notice

Preface

The communication channel,in communication acoustics,comprises of a sound source, a channel (acoustic and/or electric) and finally the receiver: the human auditory system, a complex and intricate system that forms the way sound is heard. Consequently, when developing techniques in communication acoustics, such as in speech, audio and aided hearing, it is significant to understand the time–frequency–space resolution of hearing.

This book gives an introduction to the fields which concern some kind of communication channel having the human as listener in the end; the fields together are named as "communication acoustics".

This book conveys to engineering students and researchers alike the relevant knowledge about the nature of acoustics, sound and hearing that will empower them to develop new technologies in this area from end to endgetting a thorough understanding of how sound and hearing works. It converses the multidisciplinary area of acoustics, hearing, psychoacoustics, signal processing, speech and sound quality and is appropriatefor senior undergraduate and graduate courses related to audio communication systems. It discusses the technologies and applications for sound synthesis and reproduction, and for speech and audio quality evaluation.

Table of Contents

Table of Contents

Table of Contents

Chapter 1

On the Acoustics of Emotion in Audio: What Speech, Music, and Sound have in Common

Felix Weninger[1], Florian Eyben[1], Björn W. Schuller[1,2], Marcello Mortillaro[2] and Klaus R. Scherer[2]

[1]Machine Intelligence and Signal Processing Group, Mensch-Maschine-Kommunikation, Technische Universität München, Munich, Germany
[2]Centre Interfacultaire en Sciences Affectives, Université de Genève, Geneva, Switzerland

ABSTRACT

Without doubt, there is emotional information in almost any kind of sound received by humans every day: be it the affective state of a person transmitted by means of speech; the emotion intended by a composer while writing a musical piece, or conveyed by a musician while performing it; or the affective state connected to an acoustic event occurring in the environment, in the soundtrack of a movie, or in a radio play. In the field of affective computing, there is currently some loosely connected research concerning either of these phenomena, but a holistic computational model of affect in sound is still lacking. In turn, for tomorrow's pervasive technical systems, including affective companions and robots, it is expected to be highly beneficial to understand the affective dimensions of "the sound that something makes," in order to evaluate the system's auditory environment and its own audio output. This article aims at a first step toward a holistic computational model: starting from standard acoustic feature extraction schemes in the domains of speech, music, and sound analysis, we interpret the worth of individual features across these three domains, considering four audio databases with observer annotations in the arousal and valence dimensions. In the results, we find that by selection of appropriate descriptors, cross-domain arousal, and valence regression is feasible achieving significant correlations with the observer annotations of up to 0.78 for arousal (training on sound and testing on enacted speech) and 0.60 for valence (training on enacted speech and testing on music). The high degree of cross-domain consistency in encoding the two main dimensions of affect may be attributable to the co-evolution of speech and music from multimodal affect bursts, including the integration of nature sounds for expressive effects.

INTRODUCTION

Without doubt, emotional expressivity in sound is one of the most important methods of human communication. Not only human speech, but also music and ambient sound events carry emotional information. This information is transmitted by modulation of the acoustics and decoded by the receiver – a human conversation partner, the audience of a concert, or a robot or automated dialog system. By that, the concept of emotion that we consider in this article is the one of consciously conveyed emotion (in contrast, for example, to the "true" emotion of a human related to biosignals such as heart rate). In speech, for example, a certain affective state can be transmitted through a change in vocal parameters, e.g., by adjusting fundamental frequency and loudness (Scherer et al., 2003). In music, we consider the emotion intended by the composer of a piece – and by that, the performing artist(s) as actor(s) realizing an emotional concept such as "happiness" or "sadness." This can manifest through acoustic parameters such as tempo, dynamics (forte/piano), and instrumentation (Schuller et al., 2010). In contrast to earlier research on affect recognition from singing (e.g., Daido et al., 2011), we focus on polyphonic music – by that adding the instrumentation as a major contribution to expressivity. As a connection between music and speech emotion, for example, the effect of musical training on human emotion recognition has been highlighted in related work (Nilsonne and Sundberg, 1985; Thompson et al., 2004). Lastly, also the concept of affect in sound adopted in this article is motivated by the usage of (ambient) sounds as a method of communication – to elicit an intended emotional response in the audience of a movie, radio play, or in the users of a technical system with auditory output.

In the field of affective computing, there is currently some loosely connected research concerning either of these phenomena (Schuller et al., 2011a; Drossos et al., 2012; Yang and Chen, 2012). Despite a number of perception studies suggesting overlap in the relevant acoustic parameters (e.g.,Ilie and Thompson, 2006), a holistic computational model of affect in general sound is still lacking. In turn, for tomorrow's technical systems, including affective companions and robots, it is expected to be highly beneficial to understand the affective dimensions of "the sound that something makes," in order to evaluate the system's auditory environment and its own audio output.

In order to move toward such a unified framework for affect analysis, we consider feature relevance analysis and automatic regression with respect to continuous observer ratings of the main dimensions of affect, arousal, and valence, across speech, music, and ambient sound events. Thereby, on the feature side, we restrict ourselves to non-symbolic acoustic descriptors, thus eliminating more domain-specific higher-level concepts such as linguistics, chords, or key. In particular, we use a well proven set of "low-level" acoustic descriptors for paralinguistic analysis of speech (cf. Section 2.3). Then, we address the importance of acoustic descriptors for the automatic recognition of continuous arousal and valence in a "cross-domain" setting. We show that there exist large commonalities but also strong differences in the worth of individual descriptors for emotion prediction in the various domains. Finally, we carry out experiments with automatic regression on a selected set of "generic acoustic emotion descriptors."

MATERIALS AND METHODS

Emotion Model

Let us first clarify the model of emotion employed in this article. There is a debate in the field on which type of model to adopt for emotion differentiation: discrete (categorical) or dimensional (e.g., Mortillaro et al., 2012). We believe that these approaches are highly complementary. It has been copiously shown that discrete emotions in higher dimensional space can be mapped parsimoniously into lower dimensional space. Most frequently, the two dimensions valence and arousal are chosen, although it can be shown that affective space is best structured by four dimensions – adding power and novelty to valence and arousal (Fontaine et al., 2007). Whether to choose a categorical or dimensional approach is thus dependent on the respective research context and the specific goals. Here, we chose a valence × arousal dimensional approach because of the range of affective phenomena underlying our stimuli. In addition for some of our stimulus sets only dimensional annotations were available.

Databases

Let us now start the technical discussion in this article by a brief introduction of the data sets used in the present study on arousal and valence in speech, music, and sound. The collection of emotional audio data for the purpose of automatic analysis has often been driven by computer engineering. This is particularly true for speech data – considering applications, for example, in human-computer interaction. This has led to large databases of spontaneous emotion expression, for example, emotion in child-robot interaction (Steidl, 2009) or communication with virtual humans (McKeown et al., 2012), which are however limited to specific domains. In contrast, there are data sets from controlled experiments, featuring, for example, emotions expressed ("enacted") by professional actors, with restricted linguistic content (e.g., phonetically balanced pseudo sentences) with the goal to allow for domain-independent analysis of the variation of vocal parameters (Burkhardt et al., 2005; Bänziger et al., 2012). In the case of polyphonic music, data sets are mostly collected with (commercial) software applications in mind – for example, categorization of music databases on end-user devices ("music mood recognition"; Yang and Chen, 2012). Finally, emotion analysis of general sounds has been attempted only recently (Sundaram and Schleicher, 2010; Drossos et al., 2012; Schuller et al., 2012). In this light, we selected the following databases for our analysis: the Geneva Multimodal Emotion Portrayals (GEMEP) set as an example for enacted emotional speech; the Vera am Mittag (VAM) database as an example for spontaneous emotional speech "sampled" from a "real-life" context; the "Now That's What I Call Music" (NTWICM) database for mood recognition in popular music; and the recently introduced emotional sounds database.

Enacted Emotion in Speech: the Geneva Multimodal Emotion Portrayals (GEMEP)

The GEMEP corpus is a collection of 1260 multimodal expressions of emotion enacted by 10 French-speaking actors (Bänziger et al., 2012). GEMEP comprises 18 emotions that cover all four quadrants of the arousal-valence space. The list includes the emotions most frequently used in the literature (e. g., fear, sadness, joy) as well as more subtle differentiations within emotion families (e. g., anger and irritation, fear, and anxiety). Actors expressed each emotion by using three verbal contents (two pseudo sentences and one sustained vowel) and different expression regulation strategies while they were recorded by three synchronized cameras and a separate microphone. To increase the realism and the spontaneity of the expressions, a professional director worked with the respective actor during the recording session in order to choose one scenario typical for the emotion – either by recall or mental imagery – that was personally relevant for the actor. Actors did not receive any instruction on how to express the emotion and were free to use any movement and prosody they wanted.

In the present research we consider a sub selection of 154 instances of emotional speech based on the high recognition rates reported by Bänziger et al. (2012). For this set of portrayals perceptual ratings of arousal and valence were obtained in the context of a study on the perception of multimodal emotion expressions (Mortillaro et al., unpublished). Twenty participants (10 male) listened to each of these expressions (presented in random order) and rated the content in terms of arousal and valence by using a continuous slider. Participants were given written instructions before the study. These instructions included a clear definition for each dimension that was judged. Furthermore, right before they started to rate the stimuli, they were asked whether they understood the dimensions and the two anchors and were invited to ask questions in case something was unclear. During the ratings the name of the dimension (e.g., "activation"), a brief definition (e.g., "degree of physical/physiological activation of the actor"), and the anchors ("very weak" and "very strong") were visible on the screen.

Spontaneous Emotion in Speech: the VAM Corpus

The VAM corpus (Grimm et al., 2008) was collected by the institute INT of the University Karlsruhe, Germany, and consists of audio-visual recordings taken from the German TV talk show "Vera am Mittag" (English: "Vera at noon" – Vera is the name of the talk show host). In this show, the host mainly moderates discussions between guests, e.g., by occasional questions. The corpus contains 947 spontaneous, emotionally rich utterances from 47 guests of the talk show which were recorded from unscripted and authentic discussions. There were several reasons to build the database on material from a TV talk show: there is a reasonable amount of speech from the same speakers available in each session, the spontaneous discussions between talk show guests are often rather affective, and the interpersonal communication leads to a wide variety of emotional states, depending on the topics discussed. These topics were mainly personal issues, such as friendship crises, fatherhood questions, or romantic affairs. At the time of recording, all subjects did not know that the recordings were going to be analyzed in a study of affective expression. Furthermore, the selection of the speakers was based on additional factors, such as how emotional the utterances were or which spectrum of emotions was covered by the speakers, to assure a large spectrum of different and realistic affective states. Within the VAM corpus, emotion is described in terms of three basic primitives – valence, arousal, and dominance. Valence describes the intrinsic pleasantness or unpleasantness of a situation. Arousal describes whether a stimulus puts a person into a state of increased or reduced activity. Dominance is not used for the experiments reported in this article. For annotation of the speech data, the audio recordings were manually segmented to utterance level. A large number of human annotators were used for annotation (17 for one half of the data, six for the other).

For evaluation an icon-based method that consists of an array of five images for each emotion dimension was used. Each human listener had to listen to each utterance in the database to choose an icon per emotion dimension in order to best describe the emotion heard. Afterward, the choice of the icons was mapped onto a discrete five-point scale for each dimension in the range of +1 to −1, leading to an emotion estimation (Grimm et al., 2007a).

Emotion in Music: Now that's what i Call Music (NTWICM) Database

For building the NTWICM music database the compilation "Now That's What I Call Music!" (UK series, volumes 1–69) is selected. It contains 2648 titles – roughly a week of total play time – and covers the time span from 1983 to 2010. Likewise it represents very well most music styles which are popular today; that ranges from Pop and Rock music over Rap, R&B to electronic dance music as Techno or House. While lyrics are available for 73% of the songs, in this study we only use acoustic information.

Songs were annotated as a whole, i.e., without selection of characteristic song parts. Respecting that mood perception is generally judged as highly subjective (Hu et al., 2008), four labellers were decided for. While mood may well change within a song, as change of more and less lively passages or change from sad to a positive resolution, annotation in such detail is particularly time-intensive. Yet, it is assumed that the addressed music type – mainstream popular and by that usually commercially oriented – music to be less affected by such variation as, for example, found in longer arrangements of classical music. Details on the chosen raters are provided in Schuller et al. (2011b). They were picked to form a well-balanced set spanning from rather "naïve" assessors without instrument knowledge and professional relation to "expert" assessors including a club disc jockey (DJ). The latter can thus be expected to have a good relationship to music mood, and its perception by the audiences. Further, young raters prove a good choice, as they were very well familiar with all the songs of the chosen database. They were asked to make a forced decision according to the two dimensions in the mood plane assigning values in −2, −1, 0, 1, 2 for arousal, and valence, respectively. They were further instructed to annotate according to the perceived mood, that is, the "represented" mood, not to the induced, that is, "felt" one, which could have resulted in too high labeling ambiguity. The annotation procedure is described in detail in Schuller et al. (2010), and the annotation along with the employed annotation tool are made publicly available[1].

Emotion in Sound events: Emotional Sound Database

The emotional sound database (Schuller et al., 2012)[2] is based on the on-line freely available engine FindSounds.com[3] (Rice and Bailey, 2005). It consists of 390 manually chosen sound files out of more than 10,000. To provide a set with a balanced distribution of emotional connotations, it was decided to use the following eight categories taken from FindSounds.com: *Animals, Musical instruments, Nature, Noisemaker, People, Sports, Tools,* and *Vehicles*. With this choice the database represents a broad variety of frequently occurring sounds in everyday environment. The emotional sound database was annotated by four labelers (one female, 25–28 years). They were all post graduate students working in the field of audio processing. All labelers are of Southeast-Asian origin (Chinese and Japanese), and two reported to have musical training. For the annotation these four listeners were asked to make a decision according to the two dimensions in the emotion plane assigning values on a five-point scale in {−2, −1, 0, 1, 2}for arousal and valence. They were instructed to annotate the perceived emotion and could repeatedly listen to the sounds that were presented in random order across categories. Annotation was carried out individually and independently by each of the labelers. For annotation, the procedure as described in detail inSchuller et al. (2010) was used – thus, the annotation exactly corresponds to the one used for music mood (cf. above). The annotation tool can be downloaded freely[4].

Reliability and "Gold Standard"

For all four of the databases, the individual listener annotations were averaged using the evaluator weighted estimator (EWE) as described by Grimm and Kroschel (2005). The EWE provides quasi-continuous dimensional annotations taking into account the agreement of observers. For instance *n*and dimension *d* (arousal or valence), the EWE $y_{EWE,n}^{d}$ is defined by

$$y_{EWE,n}^{d} = \frac{1}{\sum_{k=1}^{K} r_k} \sum_{k=1}^{K} r_k y_{n,k}^{d},$$

(1)

where *K* is the number of labellers, and $y_{n,k}^{d}$ is the rating of instance *n* by labeller *k* in dimension *d*. Thus, the EWE is a weighted mean rating with weights corresponding to the confidence in the labeling of rater *k* – in this study, we use the correlation coefficient r_k of rater *k*'s rating and the mean rating. By the first term in the above equation, the weights are normalized to sum up to one, in order to have the EWE in the same scale as the original ratings.

The average r_k (across the K raters) is depicted for arousal and valence annotation in the four databases in Table 1. For VAM, we observe that valence was more difficult to evaluate than arousal, while conversely, on ESD, raters agree more strongly on valence than arousal. In NTWICM, both arousal and valence have similar agreement (r = 0.70 and 0.69). Results for GEMEP are in the same order of magnitude, indicating some ambiguity despite the fact that the emotion is enacted.

Table 1: Database statistics

Database	Domain	Agreement [r]		# Annot.	# Inst.	Length [h:m]
		Arousal	Valence			
VAM	Speech (spontaneous)	0.81	0.56	6–17	947	0:50
GEMEP	Speech (enacted)	0.64	0.68	20	154	0:06
NTWICM	Music	0.70	0.69	4	2643	168:03
ESD	Sound	0.58	0.80	4	390	0:25

Furthermore, Table 1 summarizes the number of raters, number of rated instances, and length of the databases' audio. It can be seen that NTWICM is by far the largest regarding the number of instances and audio length, followed by VAM, ESD, and GEMEP. The huge differences in audio length are further due to the time unit of annotation, which is similar for VAM, ESD, and GEMEP (roughly 2–4 s of audio material), yet in NTWICM entire tracks of several minutes length of popular music were rated.

Figure 1 shows the distribution of the arousal and valence EWE ratings on the three databases considered. For the purpose of this visualization, the quasi-continuous arousal/valence ratings are discretized into five equally spaced bins spanning the interval [−1, 1] on each axis, resulting in a discretization of the arousal-valence space into 25 bins. The number of instances per bin is counted. It is evident that in VAM, instances with low valence prevail − this indicates the difficulty of creating emotionally balanced data sets by sampling audio archives. Furthermore, we observe a strong concentration of ratings in the "neutral" (center) bin of the arousal-valence space. The enacted GEMEP database is overall better balanced in terms of valence and arousal ratings − yet still, there seems to be a lack of instances with low arousal and non-neutral valence rating, although some of the chosen emotion categories (e.g., pleasure) would be expected to fit in this part. For NTWICM, we observe a concentration in the first quadrant of the valence-arousal plane, and a significant correlation between the arousal and valence ratings (Spearman's ρ = 0.61, $p \ll 0.001$). This indicates a lack of, e.g., "dramatic" music with high arousal and low valence in the chosen set of "chart" music. Finally, in ESD, ratings are distributed all over the arousal and valence scales – as shown in more detail by Schuller et al. (2012), this is due to the different sound classes in the databases having different emotional connotation (e.g., nature sounds on average being associated with higher valence than noisemakers).

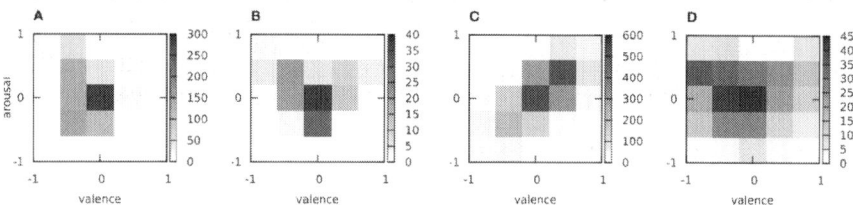

Figure 1: Distribution of valence/arousal EWE on the VAM (A), GEMEP (B), emotional sound (C), and NTWICM (D) databases: number of instances per valence/arousal bin.

Extraction of Acoustic Descriptors

In this article, the ultimate goal is automatic emotion recognition (AER) from general sound. In contrast to neighboring fields of audio signal processing such as speech or speaker recognition, which rely exclusively on rather simple spectral cues (Young et al., 2006) as acoustic features, AER typically uses a large variety of descriptors. So far no attempt has been made at defining a "standard" feature set for generic AER from sound, which may be due to the facts that AER still a rather young field with about 15 years of active research, and that emotion recognition is a multi-faceted task owing to the manifold ways of expressing emotional cues through speech, music, and sounds, and the subjective nature of the task. Some of the currently best performing approaches for automatic speech emotion recognition (Schuller et al., 2011a) use a large set of potentially relevant acoustic features and apply a large, "brute-force" set of functionals to these in order to summarize the evolution of the contours of the acoustic features over segments of typically a few seconds in length (Ververidis and Kotropoulos, 2006). This is done to capture temporal dynamics in a feature vector of fixed length and has been shown to outperform modeling of temporal dynamics on the classifier level (Schuller et al., 2009). In the process of addressing various tasks in speech and speaker characterization in a series of research challenges (Schuller et al., 2009, 2013), various large sets for the speech domain have been proposed. Little work, however, has been done on cross-domain generalization of these features, which will be the focus of the present study.

For the analysis reported on in this article, we use a well-evolved set for automatic recognition of paralinguistic phenomena – the one of the INTERSPEECH 2013 Computational Paralinguistics Evaluation baseline (Schuller et al., 2013). In this set, suprasegmental features are obtained by applying a large set of statistical functionals to acoustic low-level descriptors (cf. Tables 2 and 3). The low-level descriptors cover a broad set of descriptors from the fields of speech processing, Music Information Retrieval, and general sound analysis. For example, Mel Frequency Cepstral Coefficients (Davis and Mermelstein, 1980; Young et al., 2006) are very frequently used in ASR and speaker identification. Further, they are used in Music Information Retrieval. Spectral statistical descriptors, such as spectral variance and spectral flux, are often used in multi-media analysis, and are part of the descriptor set proposed in the MPEG-7 multimedia content description standard (Peeters, 2004). They are thus very relevant for music and sound analysis. Loudness and energy related features are obviously important for all tasks. The same holds true for the sound quality descriptors (which are used to discriminate harmonic and noise-like sounds) and the fundamental frequency and psychoacoustic sharpness. The latter is a well-known feature in sound analysis (Zwicker and Fastl, 1999). Jitter and Shimmer are micro-prosodic variations of the length and amplitudes (respectively) of the fundamental frequency for harmonic sounds. They are mainly used in voice pathology analysis, but are also good descriptors of general sound quality.

Table 2: ComParE acoustic feature set: 64 provided low-level descriptors (LLD).

	Group
4 ENERGY RELATED LLD	
Sum of auditory spectrum (loudness)	Prosodic
Sum of RASTA-style filtered auditory spectrum	Prosodic
RMS energy, zero-crossing rate	Prosodic
55 SPECTRAL LLD	
RASTA-style auditory spectrum, bands 1–26 (0–8 kHz)	Spectral
MFCC 1–14	Cepstral
Spectral energy 250–650 Hz, 1 k–4 kHz	Spectral
Spectral roll off point 0.25, 0.50, 0.75, 0.90	Spectral
Spectral flux, centroid, entropy, slope	Spectral
Psychoacoustic sharpness, harmonicity	Spectral
Spectral variance, skewness, kurtosis	Spectral
6 VOICING RELATED LLD	
F_0 (SHS and viterbi smoothing)	Prosodic
Prob. of voice	Sound quality
Log. HNR, Jitter (local, delta), Shimmer (local)	Sound quality

RESULTS

Feature Relevance

Let us now discuss the most effective acoustic features out of the above mentioned large set for single- and cross-domain emotion recognition. To this end, besides correlation coefficients (r) of features with the arousal or valence ratings, we introduce the cross-domain correlation coefficient (CDCC) as criterion. As we strive to identify features which carry similar meaning with respect to emotion in different domains, and at the same time provide high correlation with emotion in the domains by themselves, the purpose of the CDCC measure is to weigh high correlation in single domains against correlation deviations across different domains. Let us first consider a definition for two domains i and j, namely

Table 3: ComParE acoustic feature set: functionals applied to LLD contours (Table2)

	Group
FUNCTIONALS APPLIED TO LLD/Δ LLD	
Quartiles 1–3, 3 inter-quartile ranges	Percentiles
1% Percentile (≈min), 99% percentile (≈max)	Percentiles
Percentile range 1–99%	Percentiles
Position of min/max, range (max − min)	Temporal
Arithmetic mean[1], root quadratic mean	Moments
Contour centroid, flatness	Temporal
Standard deviation, skewness, kurtosis	Moments
Rel. duration LLD is above 25/50/75/90% range	Temporal
Rel. duration LLD is rising	Temporal
Rel. duration LLD has positive curvature	Temporal
Gain of linear prediction (LP), LP coefficients 1–5	Modulation
Mean, max, min, SD of segment length[2]	Temporal
FUNCTIONALS APPLIED TO LLD ONLY	
Mean value of peaks	Peaks
Mean value of peaks – arithmetic mean	Peaks
Mean/SD of inter peak distances	Peaks
Amplitude mean of peaks, of minima	Peaks
Amplitude range of peaks	Peaks
Mean/SD of rising/falling slopes	Peaks
Linear regression slope, offset, quadratic error	Regression
Quadratic regression a, b, offset, quadratic error	Regression
Percentage of non-zero frames[3]	Temporal

$$\mathrm{CDCC}^2_{f,i,j} = \frac{\left|r_f^{(i)} + r_f^{(j)}\right| - \left|r_f^{(i)} - r_f^{(j)}\right|}{2}$$

(2)

where $r_f^{(i)}$ is the correlation of feature f with the domain i, and "domain" refers to the arousal or valence annotation of a certain data set. We only consider the CDCC across the data sets (speech, music, and sound), not CDCC across arousal and valence.

It is obvious that the CDCC measure is symmetric in the sense that $\text{CDCC}^2_{f,i,j} = \text{CDCC}^2_{f,j,i}$, and that it ranges from −1 to 1. If a feature f exhibits either strong positive or strong negative correlation with both domains, the CDCC will be near one, where as it will be near −1 if a feature is strongly positively correlated with one domain yet strongly negatively correlated with the other. A CDCC near zero indicates that the feature is not significantly correlated with both domains (although it might still be correlated with either one). Thus, we can expect a regressor to show similar performance on both domains if it uses features with high CDCC.

Next, we generalize the CDCC^2 to J domains by summing up the CDCCs for domain pairs and normalizing to the range from −1 to +1,

$$\text{CDCC}^J_f = \frac{\sum_{i=1}^{J} \sum_{j=i+1}^{J} \left(\left| r_f^{(i)} + r_f^{(j)} \right| - \left| r_f^{(i)} - r_f^{(i)} \right| \right)}{J(J-1)}.$$

(3)

Intuitively, a regression function determined on features with high CDCC^J_f is expected to generalize well to all J domains.

In Tables 4 and 5, we now exemplify the CDCC^3 across the three domains on selected features, along with presenting their correlation on the individual domains. Note that for the purpose of feature selection, we treat the union of VAM and GEMEP as a single domain ("speech"). Further, in our analysis we restrict ourselves to those features that exhibit high (absolute) correlation in a single domain (termed *sound, speech,* or *music features* in the table), and those with high CDCC^3 (termed *cross-domain features*). Thereby we do not present an exhaustive list of the top features but rather a selection aiming at broad coverage of feature types. To test the significance of the correlations, we use t-tests with the null hypothesis that feature and rating are sampled from independent normal distributions. Two-sided tests are used since we are interested in discovering both negative and positive correlations. Significance levels are adjusted by Bonferroni correction, which is conservative, yet straightforward and does not require independence of the individual error probabilities.

Table 4: Cross-domain feature relevance for arousal: top features ranked by absolute correlation (r) for single domain, and CDCC across all three domains (CDCC^3).

Rank	LLD	Functional	r			CDCC^3
			Sound	Music	Speech	
SOUND FEATURES						
1	Loudness	R.q. mean	0.59**	0.16**	0.75**	0.31
4	Loudness	Lin. regr. offset	0.54**	0.27**	0.56**	0.36
6	Loudness	99-Percentile	0.53**	0.09 °	0.67**	0.23
8	Energy	R.q. mean	0.50**	0.07−	0.64**	0.21
SPEECH FEATURES						
1	Spectral flux	R.q. mean	0.38**	0.13**	0.76**	0.21
9	Δ Spectral flux	Arith. mean	0.25*	0.28**	0.68**	0.26
63	Δ MFCC 14	R.q. mean	0.14−	0.32**	0.58**	0.20
97	F0	R.q. mean	0.17−	0.09°	0.55**	0.12
MUSIC FEATURES						
1	Loudness	Mean peak dist.	0.02−	−0.58**	−0.08−	0.01
2	Spectral ent.	Mean peak dist.	0.04−	−0.54**	−0.16**	0.03
3	Loudness	Peak dist. SD	0.02−	−0.53**	−0.10−	0.02
5	MFCC 1	R.q. mean	−0.11−	−0.53**	−0.47**	0.23
CROSS-DOMAIN FEATURES						
1	Loudness	Quad. reg. offset	0.41**	0.37**	0.37**	0.37
4	Loudness	Arith. mean	0.57**	0.18**	0.73**	0.31
5	Spectral flux	Quad. reg. offset	0.32**	0.30**	0.45**	0.31
6	Δ Energy 1–4 kHz	Quartile 1	−0.32**	−0.30**	−0.59**	0.31

Table 5: Cross-domain feature relevance for valence: top features ranked by absolute correlation (r) for single domain, and CDCC across all three domains (CDCC^3)

Rank	LLD	Functional	r			CDCC^3
			Sound	Music	Speech	
SOUND FEATURES						
1	Loudness	Quartile 3	−0.31**	0.27**	−0.21**	−0.09
2	Loudness	Rise time	−0.30**	−0.21**	−0.04−	0.10
3	Loudness	R.q. mean	−0.29**	0.29**	−0.23**	−0.10
10	Spectral flux	Skewness	0.27**	−0.13**	0.11−	−0.04
SPEECH FEATURES						
1	F0	Quartile 2	0.05−	−0.07−	−0.31**	−0.01
2	Energy 1–4 kHz	Arith. mean	−0.17−	0.23**	−0.31**	−0.07
4	Δ Energy 1–4 kHz	Arith. mean	−0.08−	0.26**	−0.30**	−0.09
10	F0	Quartile 1	0.07−	−0.14**	−0.29**	0.00
MUSIC FEATURES						
1	Δ Loudness	Mean peak dist.	−0.02−	−0.65**	−0.03−	0.02
2	Loudness	Mean peak dist.	−0.12−	−0.65**	−0.04−	0.06
3	MFCC 1	Quartile 2	−0.04−	−0.61**	0.24**	−0.08
9	Spectral ent.	Mean peak dist.	0.05−	−0.57**	0.04−	−0.02
CROSS-DOMAIN FEATURES						
1	Spect. centroid	Rise time	−0.13−	−0.16**	−0.12−	0.12
2	Psy. sharpness	Rise time	−0.13−	−0.16**	−0.12−	0.12
5	Energy 250–650 Hz	IQR 1–3	−0.14−	−0.11**	−0.15*	0.12
8	MFCC 13	IQR 1–3	−0.08−	−0.20**	−0.18**	0.12

Significance denoted by **$p < 0.001$, *$p < 0.01$, °$p < 0.05$), −$p ≥ 0.05$; Bonferroni corrected p-values from two-sided paired sample t-tests.

Looking at the top sound arousal features (Table 4), we find loudness to be most relevant – in particular, the (root quadratic) mean, the linear regression offset (corresponding to a "floor value") and the 99-percentile. This is similar to the ranking for speech. Interestingly, loudness is stronger correlated than RMS energy, indicating the importance of perceptual auditory frequency weighting as performed in our loudness calculation. For music, these three loudness features are not as relevant, though still significantly correlated.

The overall best speech arousal feature is the root quadratic mean of spectral flux – indicating large differences of consecutive short-time spectra – which is interesting since it is independent of loudness and energy, which have slightly lower correlation (cf. above). The "second derivative" of the short-time spectra (arithmetic mean of Δ spectral flux) behaves in a similar fashion as spectral flux itself. However, the correlation of these features with arousal in sound and music is lower. Further, we find changes in the higher order MFCCs, such as the root quadratic mean of delta MFCC 14 to be relevant for speech and music arousal, relating to quick changes in phonetic content and timbre. Finally, mean F0, a "typical" speech feature characteristic for high arousal, is found to be relevant as expected, but does not generalize to the other domains.

The best music arousal features are related to mean peak distances – for example, in the loudness contour and the spectral entropy contour resembling occurrence of percussive instruments, indicating positive correlation between tempo and arousal. In contrast, the peak distance standard deviation is negatively correlated with arousal – thus, it seems that "periodic" pieces of music are more aroused, which can be explained by examples such as dance music. However, it seems that all these three features have a mostly musical meaning, since they only show weak correlations in sound and speech. Yet, a notable feature uniting speech and music is the (root quadratic) mean of the first MFCC, which is related to spectral skewness: arguably, a bias toward lower frequencies (high skewness) is indicative of absence of broadband (mostly percussive) instruments, and "calm" voices, and thus low arousal.

Summarizing cross-domain features for arousal, we find that the "greatest common divisor" of speech, sound, and music is loudness (and – relatedly – energy), but the behavior of functional types is interesting: the quadratic regression offset is much more relevant in the case of music than the mean loudness, which is mostly characteristic in speech and sound. In the NTWICM database of popular music, in fact we often find parabola shaped loudness contours, such that this offset indicates the intensity of the musical climax. A suitable cross-domain feature not directly related to loudness or energy is the spectral flux quadratic regression offset (the ordinate of the "high point" of spectral change).

Judging from the results in Table 5, we see that loudness is also indicative of *valence* in sound, music, and speech, but the correlations have different signs: on the one hand, loud sounds as identified by high root quadratic mean of loudness are apparently perceived as unpleasant, as are loud voices. For music, on the other hand, loudness can be indicative of high valence ("happy" music).

Among relevant speech valence features, we find mean energy (change) in the speech frequency range (1–4 kHz) and F0 (quartiles 1 and 2) – F0, however, is a "speech only" feature which exhibits low correlation in the other domains (similarly to the observations for arousal above).

Music valence features overlap with music arousal features, due to the correlation in the ratings. Among the music valence features, the median first MFCC (related to spectral skewness – cf. above) is particularly noticeable as it has "inverse" correlation on speech and music – "percussive" music with a flat spectrum is connotated with positive emotion (high valence) while "noisy" voices are characteristic of negative emotion (low valence).

Cross-domain features for valence are generally rarely significant on the individual domains and hard to interpret – here, in contrast to arousal, it seems difficult to obtain descriptors that generalize across multiple domains.

We now move from discussion of single features to a broader perspective on automatic feature selection for cross-domain emotion recognition. To this end, we consider automatically selected subsets of the ComParE feature set by the CDCC criteria. In particular, for each of arousal and valence, we choose the 200 features that show the highest CDCC[2] for the (sound, music), (sound, speech), and (music, speech) pairs of domains. Furthermore, for each of arousal and valence, we select a set of 200 features by highest CDCC[3] across all three of the sound, music, and speech domains.

In Figure 2, we summarize the obtained feature sets by the share of cepstral, prosodic, spectral, and voice quality LLDs, as well as by the share of modulation, moment, peak, percentile, regression, and temporal functionals (see Tables 2 and 3 for a list of descriptors in each of these groups). We compare the cross-domain feature sets to the full ComParE feature set as well as the "single domain" feature sets that are obtained in analogy to the cross-domain feature sets by applying the CDCC[2] to a 50% split of each corpus. A feature group is considered particularly relevant for a recognition task if its share among the selected features is larger than its share of the full feature set.

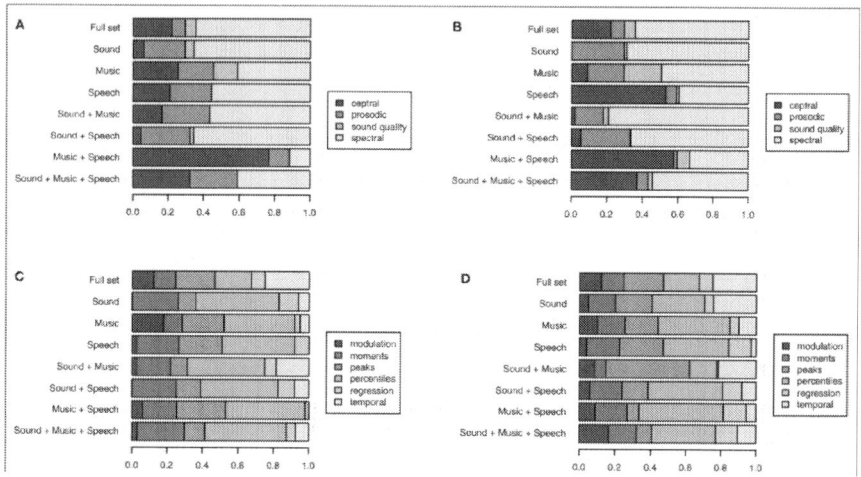

Figure 2. Feature relevance by LLD group (A: arousal, B: valence) and functional group (C: arousal, D: valence). Number of features in the top 200 features ranked by absolute correlation with gold standard for single domain and CDCC [equation (3)] for cross domain

We observe notable differences in the importance of different LLD groups; it is of particular interest for the present study to highlight the results for the considered cross-domain emotion recognition tasks: cepstral features seem to be particularly relevant for cross-domain speech and music emotion recognition. In contrast, cross-domain emotion recognition from speech and sound, and from sound and music, are dominated by "prosodic" and spectral cues such as loudness, sub-band energies, and spectral flux. Regarding relevant functional types, the summarization reveals less evident differences between the tasks; still, percentile type functionals seem to be particularly promising for all of the tasks considered.

Automatic Classification Experiments

Finally, we demonstrate the predictive power of the obtained cross-domain feature sets in automatic regression. In automatic regression, the parameters of a regression function on N-dimensional feature vectors are optimized to model the assignment of L "learning" vectors (e.g., feature vectors of emotional utterances) to the gold standard (e.g., the arousal observer rating). Then, the regression function is evaluated on a disjoint set of test vectors and the correlation of the function's predictions and the test set gold standard is computed as a measure of how well the regression function generalizes to "unseen" test data. In the present study, it is of particular interest to consider cross-domain evaluation, i.e., training on data from one domain (e.g., enacted speech) and evaluating on another domain (e.g., sound). In this context, we also treat spontaneous and enacted speech as different domains, as such analysis is receiving increasing attention at the moment (Bone et al., 2012) also due to practical reasons: for instance, it is of interest to determine if training on "prototypical" data from a controlled experiment (such as the GEMEP database) can improve automatic emotion recognizers applied "in the wild," e.g., to media analysis (such as given by the VAM database). For reference, we also consider within-domain regression in a twofold cross-validation manner.

For each learning set, we determine a multivariate linear regression function by means of support vector regression (SVR) (Smola and Schölkopf, 2004), which defines a real valued mapping

$$f(\mathbf{x}) = \mathbf{w}^T \mathbf{x} + b$$

(4)

of N-dimensional feature vectors $\mathbf{x}$ to a regression value $f(\mathbf{x})$. $\mathbf{w}$ is the normal vector of the N-dimensional hyperplane describing the regression function, and b is a scalar offset. Specifically for SVR, the primary optimization goal is *flatness* of the regression function, which is defined as low norm of the weight vector $\mathbf{w}$. This is related to the notion of *sparsity* and crucial to avoid over-fitting of the model parameters in the present case of high dimensional feature spaces. The trade-off between flatness of the weight vector and deviation of the regression values from the gold standard on the learning set is modeled as a free parameter C in the optimization (cf. Smola and Schölkopf, 2004 for details). In our experiments, C is set to 10^{-5} for within-domain regression and 10^{-5} for cross-domain regression. The optimization problem is solved by the frequently used Sequential Minimal Optimization algorithm (Platt, 1999). To foster reproducibility of our research, we use the open-source machine learning toolkit Weka (Hall et al., 2009). Unsupervised mean and variance normalization of each feature per database is applied since SVR is sensitive to feature scaling.

In Table 6, the correlation coefficients (r) of automatic within-domain and cross-domain regression with the arousal observer ratings are displayed. First, we consider regression using the full 6373-dimensional ComParE feature set. In within-domain regression, results ranging from $r = 0.54$ (sound) up to $r = 0.85$ (enacted speech) are obtained, which are comparable to previously obtained results on sound, music, and spontaneous speech (Grimm et al., 2007b; Schuller et al., 2011b,2012). Especially the result for music is notable, since we do not use any "hand-crafted" music features such as chords or tempo. In cross-domain regression, significant correlations are obtained except for the case of training on music and evaluating on sound. However, the mean r across all training and testing conditions (0.50) is rather low.

Considering automatic feature selection by CDCC[2] for each combination of two domains, results in Table 6B indicate a drastic gain in performance especially for cross-domain regression. However, also the results in within-domain regression are improved. All correlations are significant at the 0.1% level. Particularly, using CDCC based feature selection robust regression (achieving $r > 0.76$) is possible across enacted and spontaneous speech. Further, it is notable that the average result across the four testing databases does not vary much depending on the training database used, indicating good generalization capability of the selected features. The overall mean r in this scenario is 0.65.

Finally, if we select the top features by CDCC[3] on all databases (treating speech as a single domain for the purpose of feature selection), it is notable that we still obtain reasonable results (mean r of 0.58) despite the fact that the top features by CDCC[3] exhibit comparably low correlation with the target labels on the single domains (cf. Table 4).

Summarizing the results for valence regression (Table 7), we observe that using the full feature set, we cannot obtain reasonable results in cross-domain regression. Among the cross-domain results, the only significant positive correlations are obtained in evaluation on spontaneous speech, however, these are lower than the correlation of the single best speech

Table 6: Results of within-domain and pair-wise cross-domain support vector regression on arousal observer ratings for sound (emotional sound database), music (NTWICM database), and spontaneous and enacted speech (VAM/GEMEP databases)

r	Test on				Mean
Train on	Sound	Music	Speech		
			Sp.	En.	
(A) FULL FEATURE SET					
Sound	0.54**	0.14**	0.70**	0.64**	0.51
Music	0.11−	0.65**	0.46**	0.39**	0.40
Speech/Sp.	0.38**	0.37**	0.81**	0.80**	0.59
Speech/En.	0.20*	0.32**	0.60**	0.85**	0.49
Mean	0.30	0.37	0.64	0.67	0.50
(B) 200 TASK-SPECIFIC FEATURES					
Sound	0.59**	0.46**	0.76**	0.79**	0.65
Music	0.46**	0.67**	0.73**	0.75**	0.65
Speech/Sp.	0.54**	0.47**	0.83**	0.78**	0.66
Speech/En.	0.56**	0.46**	0.77**	0.85**	0.66
Mean	0.54	0.52	0.77	0.79	0.65
(C) 200 GENERIC FEATURES					
Sound	0.56**	0.35**	0.78**	0.56**	0.56
Music	0.38**	0.66**	0.74**	0.63**	0.60
Speech/Sp.	0.44**	0.43**	0.82**	0.69**	0.59
Speech/En.	0.31**	0.45**	0.77**	0.78**	0.58
Mean	0.42	0.47	0.78	0.67	0.58

*Significance denoted by **p < 0.001, *p < 0.01, −p ≥ 0.05; Bonferroni corrected p-values from two-sided paired sample t-tests. Full ComParE feature set (cf. **Tables 2** and **3**); 200 top features selected by $CDCC^2$ for specific within-domain or cross-domain regression tasks; Generic features: 200 features selected by $CDCC^3$ across sound, music, and speech domains (cf. **Table 4**).*

features. Interestingly, we observe significant negative correlations when evaluating on music and training on another domain, which is consistent with the fact that some of the music valence features are "inversely" correlated with the target label in the other domains (cf., e.g., the discussion of median MFCC 1 above). In the within-domain setting, it can be observed that regression on valence in music is possible with high robustness ($r = 0.80$). This is all the more noticeable since this correlation is higher than the one obtained in arousal regression, while for the other domains, valence seems to be harder to recognize than arousal. This can partly be attributed to the fact that in the analyzed music data, the valence rating is correlated to the arousal rating.

Table 7: Results of within-domain and pair-wise cross-domain support vector regression on valence observer ratings for sound (emotional sound database), music (NTWICM database), and spontaneous and enacted speech (VAM/GEMEP databases)

r		Test on			Mean
Train on	**Sound**	**Music**	**Speech**		
			Sp.	**En.**	
(A) FULL FEATURE SET					
Sound	0.40**	−0.11**	0.21**	−0.02⁻	0.12
Music	−0.17 °	0.80**	−0.13*	0.08⁻	0.15
Speech/Sp.	0.11⁻	−0.15**	0.46**	0.21⁻	0.16
Speech/En.	−0.06⁻	−0.18**	0.12*	0.26°	0.03
Mean	0.07	0.09	0.17	0.13	0.12
(B) 200 TASK-SPECIFIC FEATURES					
Sound	0.51**	0.36**	0.27**	0.48**	0.41
Music	0.40**	0.82**	0.33**	0.52**	0.52
Speech/Sp.	0.30**	0.45**	0.44**	0.26°	0.36
Speech/En.	0.45**	0.60**	0.36**	0.50**	0.48
Mean	0.41	0.56	0.35	0.44	0.44
(C) 200 GENERIC FEATURES					
Sound	0.26**	0.41**	0.27**	0.12⁻	0.27
Music	0.27**	0.75**	0.33**	0.25°	0.40
Speech/Sp.	0.20*	0.45**	0.35**	0.19⁻	0.30
Speech/En.	0.20**	0.44**	0.32**	0.23⁻	0.30
Mean	0.23	0.52	0.32	0.20	0.32

*Significance denoted by **$p < 0.001$, *$p < 0.01$, °$p < 0.05$, ⁻$p \geq 0.05$; Bonferroni corrected p-values from two-sided paired sample t-tests. Full ComParE feature set (cf. **Tables 2** and **3**); 200 top features selected by CDCC² for specific within-domain or cross-domain regression tasks; Generic features: 200 features selected by CDCC³ across sound, music, and speech domains (cf. **Table 5**).*

Concerning feature selection by $CDCC^2$ (Table 7B), we observe a boost in the obtained correlations (mean = 0.44, compared to 0.12 without feature selection). For instance, when training on enacted speech and evaluating on music, we obtain a significant r of 0.60. This result is interesting in so far as the best selected feature for this particular cross-domain setting, namely the flatness of the loudness contour, only exhibits a correlation of 0.28, respectively 0.27, with the valence rating on the NTWICM (music) and GEMEP (enacted speech) databases. Thus, the 200 $CDCC^2$-selected features for this regression task seem to be of complementary nature. Furthermore, by applying feature selection in the within-domain setting, best results are obtained for sound (r = 0.51), music (r = 0.82), and enacted speech (r = 0.50) valence recognition. However, regarding the issue of enacted vs. spontaneous speech, we find that regressors trained on one type do not generalize well to the other, which is in contrast to the finding for arousal.

Finally, when applying the "generic valence feature set" obtained from the $CDCC^3$ ranking across sound, music, and speech, we obtain an average correlation of 0.32. Results are considerably below the $CDCC^2$ results particularly for sound and enacted speech. This – again – points at the difficulty of finding features that generalize to valence recognition across domains. However, it is notable that robust results (r = 0.75) are obtained in within-domain music recognition using the generic feature set, of which the "best" feature (rise time of spectral centroid) only has an (absolute) correlation of 0.16 with the music valence rating.

DISCUSSION

We have presented a set of acoustic descriptors for emotion recognition from audio in three major domains: speech (enacted and spontaneous), music, and general sound events. Using these features, we have obtained notable performances in within-domain regression – particularly, these surpass the so far best published results on the NTWICM database (Schuller et al., 2011b) despite the fact that the latter study used hand-crafted music features rather than the generic approach pursued in the present paper.

We have found that it is rather hard to obtain features that are equally well correlated across the three domains. For arousal, such features comprise mostly loudness-related ones. In contrast, we have not been able to obtain features that are significantly correlated with the valence rating in all domains. A further notable result for valence is that some features have an "inverse" meaning in different domains (i.e., significant correlations with different signum), while this does not occur for arousal. It will be subject of further research whether this is simply due to the correlation of intended arousal and valence in popular music or to more fundamental differences.

This phenomenon has motivated the introduction of a "cross-domain correlation coefficient" which summarizes the differences in correlation across multiple domains. Using this coefficient, we were able to provide an automatic method of selecting generalizing features for cross-domain arousal and valence recognition. In the result, cross-domain arousal and valence regression has been proven feasible, achieving significant correlations with the observer annotations.

The degree of cross-domain consistency in encoding the two main dimensions of affect – valence and arousal – demonstrated in this article is quite astounding. Music has often been referred to as the "language of emotion" and a comprehensive review of empirical studies on the expression of emotion in speech and music (Juslin and Laukka, 2003) has confirmed the hypothesis that the acoustic parameters marking certain emotions are quite similar in music and speech (cf. also Ilie and Thompson, 2006). Scherer (1991) has suggested that speech and emotion may have evolved on the basis of primitive affect bursts serving similar communicative functions across many mammalian species. Ethological work shows that expression and impression are closely linked, suggesting that, in the process of conventionalization and ritualization, expressive signals may have been shaped by the constraints of transmission characteristics, limitations of sensory organs, or other factors. The resulting flexibility of the communication code is likely to have fostered the evolution of more abstract, symbolic language, and music systems, in close conjunction with the evolution of the brain to serve the needs of social bonding and efficient group communication.

As vocalization, which remained a major modality for analog emotion expression, became the production system for the highly formalized, segmental systems of language and singing, both of these functions needed to be served at the same time. Thus, in speech, changes in fundamental frequency (F0), formant structure, or characteristics of the glottal source spectrum can, depending on the language and the context, serve to communicate phonological contrasts, syntactic choices, pragmatic meaning, or emotional expression. Similarly, in music, melody, harmonic structure, or timing may reflect the composer's intentions, depending on specific traditions of music, and may simultaneously induce strong emotional moods. This fusion of two signal systems, which are quite different in function and in structure, into a single underlying production mechanism, vocalization, has proven to be singularly efficient for the purpose of communication, and the relatively high degree of convergence as demonstrated by the correlations found in our study suggests that it might be possible to identify elements of a common code for emotion signaling. Recently, Scherer (2013) has reviewed theoretical proposals and empirical evidence in the literature that help to establish the plausibility of this claim, in particular, the evolutionary continuity of affect vocalizations, showing that anatomical structures for complex vocalizations existed before the evidence for the presence of representational systems such as language.

As to the cross-domain consistency with different kinds of environmental sounds, it seems quite plausible to assume that once speech and music were decoupled from actually occurring affect bursts and took on representational functions, different kinds of nature sounds were used in speech and music both for reference to external events and expressive functions. It seems reasonable to assume that the type of representational coding was informed by the prior, psychobiological affect code, particularly with respect to the fundamental affect dimensions of valence and arousal.

Empirical studies like the one reported here, using machine learning approaches, may complement other approaches to examine the evolutionary history of affect expression in speech and music by empirically examining, using large corpora of different kinds of sound events, the extent to which auditory domains exhibit cross-domain consistency and which common patterns are particularly frequent.

ACKNOWLEDGMENTS

This study has received funding from the European Commission (grant no. 289021, ASC-Inclusion).

REFERENCES

1. Bänziger, T., Mortillaro, M., and Scherer, K. R. (2012). Introducing the Geneva multimodal expression corpus for experimental research on emotion perception. *Emotion* 12, 1161–1179. doi:10.1037/a0025827

2. Bone, D., Lee, C.-C., and Narayanan, S. (2012). "A robust unsupervised arousal rating framework using prosody with cross-corpora evaluation," in *Proceeding of the Interspeech* (Portland, OR: ISCA).

3. Burkhardt, F., Paeschke, A., Rolfes, M., Sendlmeier, W., and Weiss, B. (2005). "A database of German emotional speech," in *Proceeding of the Interspeech* (Lisbon: ISCA), 1517–1520.

4. Daido, R., Hahm, S., Ito, M., Makino, S., and Ito, A. (2011). "A system for evaluating singing enthusiasm for karaoke," in *Proceeding of the ISMIR* (Miami, FL: International Society for Music Information Retrieval), 31–36.

5. Davis, S. B., and Mermelstein, P. (1980). Comparison of parametric representations for monosyllabic word recognition in continuously spoken sentences. *IEEE Trans. Acoust.* 28, 357–366. doi:10.1109/TASSP.1980.1163420

6. Drossos, K., Floros, A., and Kanellopoulos, N.-G. (2012). "Affective acoustic ecology: towards emotionally enhanced sound events," in *Proceedings of the 7th Audio Mostly Conference: A Conference on Interaction with Sound* (New York, NY: ACM), 109–116.

7. Fontaine, J., Scherer, K. R., Roesch, E., and Ellsworth, P. (2007). The world of emotion is not two-dimensional. *Psychol. Sci.* 18, 1050–1057. doi:10.1111/j.1467-9280.2007.02024.x

8. Grimm, M., and Kroschel, K. (2005). "Evaluation of natural emotions using self assessment manikins," in *Proceeding of the ASRU* (Cancún: IEEE), 381–385.

9. Grimm, M., Kroschel, K., Mower, E., and Narayanan, S. (2007a). Primitives-based evaluation and estimation of emotions in speech. *Speech Commun.* 49, 787–800. doi:10.1016/j.specom.2007.01.010

10. Grimm, M., Kroschel, K., and Narayanan, S. (2007b). "Support vector regression for automatic recognition of spontaneous emotions in speech," in *Proceeding of the International Conference on Acoustics, Speech and Signal Processing (ICASSP)*, Vol. IV (Honolulu, HI: IEEE), 1085–1088.

11. Grimm, M., Kroschel, K., and Narayanan, S. (2008). "The Vera am Mittag German audio-visual emotional speech database," in *Proceeding of the IEEE International Conference on Multimedia and Expo (ICME)* (Hannover: IEEE), 865–868.

12. Hall, M., Frank, E., Holmes, G., Pfahringer, B., Reutemann, P., and Witten, I. H. (2009). The WEKA data mining software: an update. *SIGKDD Explor.* 11, 10–18. doi:10.1145/1656274.1656278

13. Hu, X., Downie, J. S., Laurier, C., Bay, M., and Ehmann, A. F. (2008). "The 2007 MIREX audio mood classification task: lessons learned," in *Proceeding of the ISMIR* (Philadelphia: International Society for Music Information Retrieval), 462–467.
14. Ilie, G., and Thompson, W. F. (2006). A comparison of acoustic cues in music and speech for three dimensions of affect. *Music Percept.* 23, 319–329. doi:10.1525/mp.2006.23.4.319
15. Juslin, P. N., and Laukka, P. (2003). Communication of emotions in vocal expression and music performance: different channels, same code? *Psychol. Bull.* 129, 770–814. doi:10.1037/0033-2909.129.5.770
16. McKeown, G., Valstar, M., Cowie, R., Pantic, M., and Schröder, M. (2012). The SEMAINE database: annotated multimodal records of emotionally colored conversations between a person and a limited agent. *IEEE Trans. Affect. Comput.* 3, 5–17. doi:10.1109/T-AFFC.2011.20
17. Mortillaro, M., Meuleman, B., and Scherer, K. R. (2012). Advocating a componential appraisal model to guide emotion recognition. *Int. J. Synth. Emot.* 3, 18–32. doi:10.4018/jse.2012010102
18. Nilsonne, A., and Sundberg, J. (1985). Differences in ability of musicians and nonmusicians to judge emotional state from the fundamental frequency of voice samples. *Music Percept.* 2, 507–516. doi:10.2307/40285316
19. Peeters, G. (2004). *A Large Set of Audio Features for Sound Description.* Technical Report. Paris: IRCAM.
20. Platt, J. C. (1999). "Fast training of support vector machines using sequential minimal optimization," in *Advances in Kernel Methods: Support Vector Learning* (Cambridge, MA: MIT Press), 185–208.
21. Rice, S. V., and Bailey, S. M. (2005). "A web search engine for sound effects," in *Proceeding of the 119th Conference of the Audio Engineering Society (AES)* (New York: Audio Engineering Society).
22. Scherer, K. R. (1991). "Emotion expression in speech and music," in *Music, Language, Speech, and Brain*, eds J. Sundberg, L. Nord, and R. Carlson (London: Macmillan), 146–156.
23. Scherer, K. R. (2013). "Emotion in action, interaction, music, and speech," in *Language, Music, and the Brain: A Mysterious Relationship*, ed. M. Arbib (Cambridge, MA: MIT Press), 107–139.
24. Scherer, K. R., Johnstone, T., and Klasmeyer, G. (2003). "Vocal expression of emotion," in *Handbook of Affective Sciences*, eds R. J. Davidson, K. R. Scherer, and H. H. Goldsmith (Oxford, NY: Oxford University Press), 433–456.
25. Schuller, B., Batliner, A., Steidl, S., and Seppi, D. (2011a). Recognising realistic emotions and affect in speech: state of the art and lessons learnt from the first challenge. *Speech Commun.* 53, 1062–1087. doi:10.1016/j.specom.2011.01.011

26. Schuller, B., Weninger, F., and Dorfner, J. (2011b). "Multi-modal non-prototypical music mood analysis in continuous space: reliability and performances," in *Proceedings 12th International Society for Music Information Retrieval Conference, ISMIR 2011* (Miami, FL: ISMIR), 759–764.

27. Schuller, B., Dorfner, J., and Rigoll, G. (2010). Determination of nonprototypical valence and arousal in popular music: features and performances. *EURASIP J. Audio Speech Music Process.* 2010:735854. doi:10.1186/1687-4722-2010-735854

28. Schuller, B., Hantke, S., Weninger, F., Han, W., Zhang, Z., and Narayanan, S. (2012). "Automatic recognition of emotion evoked by general sound events," in *Proceedings 37th IEEE International Conference on Acoustics, Speech, and Signal Processing, ICASSP 2012* (Kyoto: IEEE), 341–344.

29. Schuller, B., Steidl, S., and Batliner, A. (2009). "The INTERSPEECH 2009 emotion challenge," in *Proceeding of the Interspeech* (Brighton: ISCA), 312–315.

30. Schuller, B., Steidl, S., Batliner, A., Vinciarelli, A., Scherer, K., Ringeval, F., et al. (2013). "The INTERSPEECH 2013 computational paralinguistics challenge: social signals, conflict, emotion, autism," in *Proceedings INTERSPEECH 2013, 14th Annual Conference of the International Speech Communication Association* (Lyon: ISCA).

31. Smola, A., and Schölkopf, B. (2004). A tutorial on support vector regression. *Stat. Comput.* 14, 199–222. doi:10.1023/B:STCO.0000035301.49549.88

32. Steidl, S. (2009). *Automatic Classification of Emotion-Related User States in Spontaneous Children's Speech.* Berlin: Logos Verlag.

33. Sundaram, S., and Schleicher, R. (2010). "Towards evaluation of example-based audio retrieval system using affective dimensions," in *Proceeding of the ICME* (Singapore: IEEE), 573–577.

34. Thompson, W. F., Schellenberg, E. G., and Husain, G. (2004). Decoding speech prosody: do music lessons help? *Emotion* 4, 46–64. doi:10.1037/1528-3542.4.1.46

35. Ververidis, D., and Kotropoulos, C. (2006). Emotional speech recognition: resources, features, and methods. *Speech Commun.* 48, 1162–1181. doi:10.1016/j.specom.2006.04.003

36. Yang, Y.-H., and Chen, H.-H. (2012). Machine recognition of music emotion: a review. *ACM Trans. Intell. Syst. Technol.* 3, 1–30. doi:10.1145/2168752.2168754

37. Young, S. J., Evermann, G., Gales, M. J. F., Hain, T., Kershaw, D., Liu, X., et al. (2006). *The HTK Book, Version 3.4.1.* Cambridge: Cambridge University Engineering Department.

CITATION

FelixWeninger, Florian Eyben, BjörnW. Schuller, Marcello Mortillaro and Klaus R. Scherer, On the acoustics of emotion in audio: what speech, music, and sound have in common, doi.org/10.3389/fpsyg.2013.00292

Chapter 2

Acoustics and Vibro-Acoustics Applied in Space Industry

Rogério Pirk[1], Carlos d'Andrade Souto[1], Gustavo Paulinelli Guimarães[1] and Luiz Carlos Sandoval Góes[2]

[1]Institute of Aeronautics and Space (IAE) and Technological Institute of Aeronautics (ITA), São José dos Campos, Brazil
[2]Technological Institute of Aeronautics (ITA), São José dos Campos, Brazil

INTRODUCTION

During flight, Expandable Launch Vehicles (ELV) are excited by severe acoustic loads in three phases of flight: lift off, transonic flight and maximum dynamic pressure instant [1]. As such, principles to make onboard equipment compatible with the mission environments must be adopted. At lift off, the highly intense acoustic loads occur; and these levels are usually adopted to qualify payloads and equipments. However, during the transonic flight and maximum dynamic pressure phase, acoustic excitation is also present and such characteristics are also significant for performance evaluation as well as for specific system dynamic qualification/acceptance programs. In this way, noise control treatments (NCT) shall be adopted to alleviate internal vibro-acoustic environments, in view of decreasing costs and developments.

The hostile in-flight environments can damage sensors/conditioners as well as make measurements unreliable. In this way, installation adapters must be designed to protect the sensors. The acoustics of such protective cavities influence the measured sound pressure level (SPL) As such, the cavities must be analyzed and their amplitude-frequency characteristics evaluated. Finally, the measurement corrections, necessary to obtain the actual external SPL, are determined.

Concerning the internal environment found during flights, important launcher subsystems as payload fairing (PLF) and equipment bays shall be investigated and vibro-acoustic analysis can be done, as pointed by [2], [3] and [4]. The PLF is the structural compartment of a launcher where the payload is placed during the flight mission. PLF inner acoustics and its attenuation designs, using virtual prototypes are analyzed using deterministic and statistic techniques. However, when in-flight loading are not characterized, the accounted external air-borne excitation can be those described in [5]. In a similar way, SPL along the launcher structure at lift-off can also be estimated [6]. Furthermore, an alternative procedure to characterize external SPL during flight can also be adopted as described by [7]. Passive vibration control techniques can be used to attenuate structure-borne vibration and the use of viscoelastic materials adding structural damping to reduce the magnitude of vibrations is a well-known solution, usually applied in space and aeronautical industries. On the other hand, the use of active vibration control (AVC) is still considered difficult to be implemented in space industry.

For acoustic noise attenuation, the standard practice is to use passive techniques like blankets ([8] and [9]), which attenuate sound by trapping the energy in the blanket material and dissipating it as heat [10] and Helmholtz resonators tuned to absorb acoustic energy at one or some specific frequencies, typically the cavity frequencies as done by [11].

Another acoustic crucial subject in space industry is combustion instability, since it can severely impair the operation of Liquid Propelled Rocket Engine (LPRE) [12]. In this way, solutions for instability problems in combustion chambers of LPRE as well as solid rocket motors (SRM) are of large interest. In [13], it is described that combustion instability can be verified when the power spectrum of the acoustic pressure measured during tests is analyzed. When an oscillation is observed, i.e., combustion instability, well-defined sound pressure peaks, summed to the background noise are present. Such peaks are correlated to the resonance frequencies of the combustion chamber. In this way, the coupling of acoustic natural frequencies and burning oscillations of the combustion chamber occurs, which can cause instabilities and consequent unexpected behavior as efficiency loss or even the explosion of the engine. In the early developing phases of liquid rocket engines, it is usually proposed the investigation of different combustion chamber configurations [14]. This is usually done in two steps as follows: using theoretical calculation and through experimental measurements. In this way, theoretical and experimental natural frequencies of the acoustic cavity are obtained. Further studies must be performed, applying devices and techniques to attenuate pressure oscillations inside combustion chambers and devices as Helmholtz Resonators, baffles and ¼ wave filters are largely used [15] and [16].

This chapter describes three case studies applied on space industry. Firstly, analytical and numerical modelings of in-flight external microphone protection devices are described. Testing procedures and the SPL measurement correction factors determination are also presented. As a second case study, deterministic and statistical coupled vibro-acoustic analysis techniques are used to estimate PLF internal SPL at lift-off as well as to assess the effect of including NCT (blanket materials) on its skin. The modeling procedures and experimental ground test are described. Finally, in the third case, the acoustic characterization of combustion chambers is presented. Cold tests are described as well as the theoretical modeling procedures. The pressure attenuation control technique using Helmholtz Resonators are also presented. In all three case studies, theoretical x experimental results are depicted.

EXTERNAL ON BOARD MICROPHONES INSTALLATION DEVICES

At lift off, the source of the acoustic noise is the gas stream ejected by the motors (Fig. 1). Such acoustic pressure lies in the range of 140 to 180 dB near the rocket and is very close to an acoustic diffuse field (ADF) noise. At transonic flight the launcher is excited by the turbulent boundary layer (TBL) in the neighborhood of the shock waves. According to [17], when the maximum dynamic pressure occurs, the unsteady pressure field applied to the launcher is due to aerodynamic noise. The characteristics of such noise are very different from those at lift off. Non-attached flows increase the pressure in low-frequencies, which excite the launcher first structural modes. A simulation of the VLS-1 flight aerodynamics was done by Academician V. P. Makeyev State Rocket Centre (SRC- Makeyev), as shown in Fig. 2. Notice that the upper nose and 1st stage noses are the most exposed regions to aero-acoustic noise.

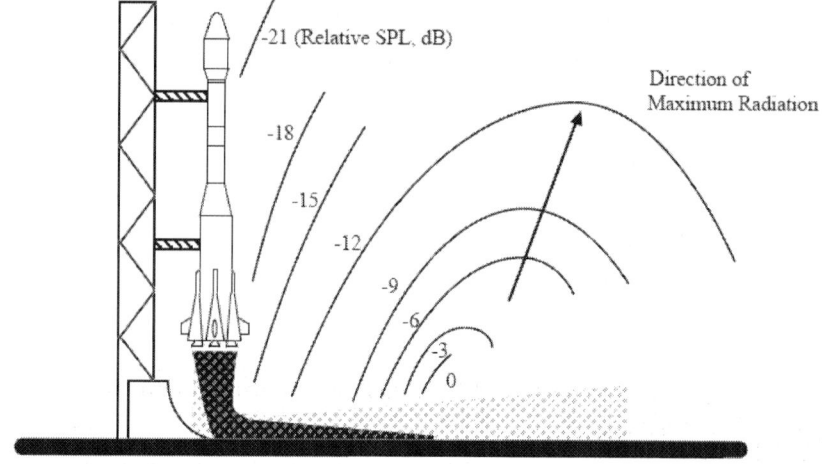

Figure 1. Acoustic noise at lift off

Figure 2. In-flight aero-acoustic noise

In view of having a good characterization of the in-flight acoustic loads acting upon the launcher structure, external acoustic measurements are required. Due to the high SPL and hostile environments found during flight, special microphones and adapters are specified. Such adapters must be designed in order to provide appropriate microphone/pre-amplifier installation and protection. Besides, when necessary, measurement correction procedures must be adopted. In this way, measurement programs for in-flight external acoustic characterization shall be developed, which may take into account three main phases as: preparation for experimental studies and acoustic testing sensors, ground development testing of acoustic sensors and methodology for reading acoustic pressures during flight.

Three different adapters for ¼'' microphones were conceived, as described by [18]. On the upper parts of the launcher, two different configurations can be adopted. Firstly, at the PLF, heating and propellant dust effects are not significant; therefore, a structure flush installation (Fig. 3a) can be used. In this case, the measured SPL can be read directly. The second configuration, straight adapter (Fig. 3b), is applied for microphones installed near the equipment bays, where one has temperature and dust influences and, therefore, the sensor/conditioner must be protected. For such an assemblage, the protection channel dynamics directly affect the sensor response and, as a result, a measurement correction must be done.

On the bottom, the intense SPL at lift off generate a severe acoustic excitation of the first stage back modules region. Highly hostile dust, hot gas flow, heat flux and temperature environments are present during the motors operations. Nevertheless, the angular adapter must be used to install acoustic microphone/pre-amplifier, as shown in Fig. 3c. Notice in Figs. 3b and 3c, that the adapters were designed with small acoustic straight and angular cavities, respectively. When acoustically excited, the acoustical responses of such cavities directly influence the measured SPL, since the external pressure excitation profile and the measured signal are related by the cavity transfer function.

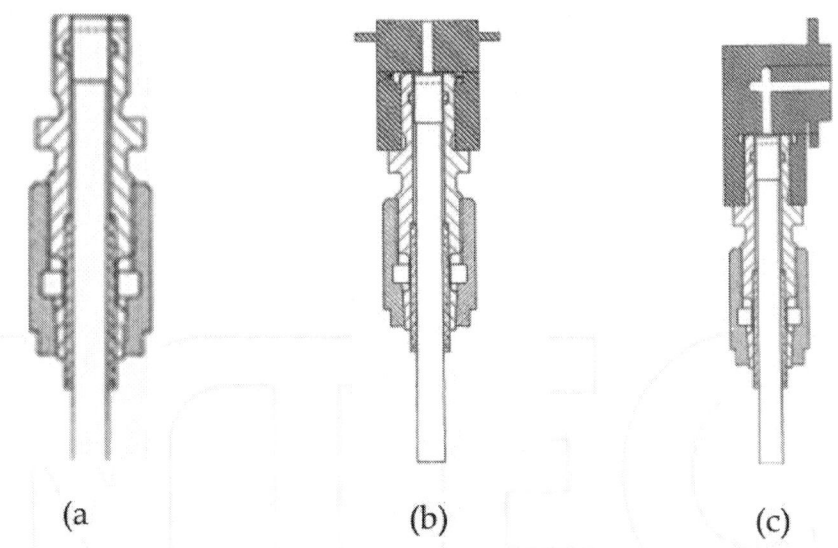

(a)   (b)   (c)

Figure 3. a) Flush adapter; (b) Straight adapter; (c) Angular adapter

In order to determine straight and angular adapters' dynamics, analytical and numerical calculations are done. The transfer functions of these channels are evaluated during ground acoustic tests, on which acoustic excitation with SPL close to those expected during flight is used to excite the cavities. Consequently, the measured SPL as well as the channels transfer functions are determined. Finally, the measurement corrections are determined, which may be applied when these adapters are used.

Mathematical Models

Analytical Model
In view of describing the dynamical behavior of the protective channels, one can assume the straight and angular channels as Helmholtz Resonators, which the channels and the space for microphone installation are accounted as the resonator throat and volume, as described by Eq. (1) [19].

$$f_0 = \frac{c_0}{2\pi}\sqrt{\frac{S}{V_0(l+l_c)}}$$

(1)

where:f0: natural frequency,c0: sound speed,S: cross section of the resonator throat,V0: volume of the resonator cavity,l: length of the resonator throat,lc=0.8r: end correction andr: radius of the resonator throat cross section.

By considering the dimensions of the adapters into Eq. (1), one can calculate the natural frequencies of the straight and angular channels shown in Figs. 3b and 3c. These calculations are the starting point to assess the accuracy of the numerical models, built by using the Finite Element Method, once analytical x numerical natural frequencies can be compared.

Numerical Model by Finite Element Method (Fem)
In a similar way as in structural dynamics, an acoustic cavity FEM model will have an acoustic stiffness matrix [Ka], an acoustic mass matrix [M_a], acoustic excitation vectors {F_{ai}} and an acoustic damping matrix [C_a]. The combination of these components yields the acoustic finite element model, which can be solved for the unknown nodal pressure values pi [20].

$$\left([K_a] + j\omega[C_a] - \omega^2[M_a]\right).\{p_i\} = \{F_{ai}\}$$

(2)

Acoustic finite element models of the three adapters cavities are built. All cavities' surfaces were considered as rigid walls but the openings that are in direct contact with the external acoustic environment. In such cases, opened surfaces approximated using prescribed nodal pressures (equal to 0 for the eigenanalysis) were considered. The fluid inside the cavities is assumed as air at 15° C (c=340 m/s, ρ=1.225 Kg/m^3, values used in this entire chapter).

Linear tetrahedral fluid elements are used in all three meshes. In order to have good prediction accuracy in the frequency range of interest, the general rule of thumb that requires at least 6 elements per wavelength is adopted. The main meshing characteristics are described in Table 1.

The acoustic load generated at lift off is simulated as an acoustic diffuse field (ADF). According to [21], an ADF is defined as an acoustic field in which the SPL is equal at any location and have an identical energy distribution in all directions. Such ADF can be obtained in an acoustic reverberant chamber, where the reflections along the rigid walls lead to this field. A formal way to describe an ADF consists on superimposing an infinite number of uncorrelated plane waves through different directions. In a FEM model, a finite number of uncorrelated plane waves can be generated and the pressure due the superposition of all the uncorrelated plane waves can then be applied as prescribed nodal pressures on the cavity's open surface (see [22]).

Table 1. Meshes data

Parameters	Mesh		
	Flush adapter	Straight adapter	Angular adapter
Number of elements	4,419	3,986	10,266
Number of nodes	1,009	980	2,367
Maximum frequency (6 elements/wavelength)	45,374 Hz	44,593 Hz	44,199 Hz

The FEM models of the three adapters are shown in Figs. 4a, 4b and 4c. The cavity's transfer functions are calculated by imposing prescribed nodal pressures in the nodes marked with small green arrows. In order to save time and computational efforts, a modal solution method is adopted using the first 14 modes. A modal damping of 5% is considered in these calculations.

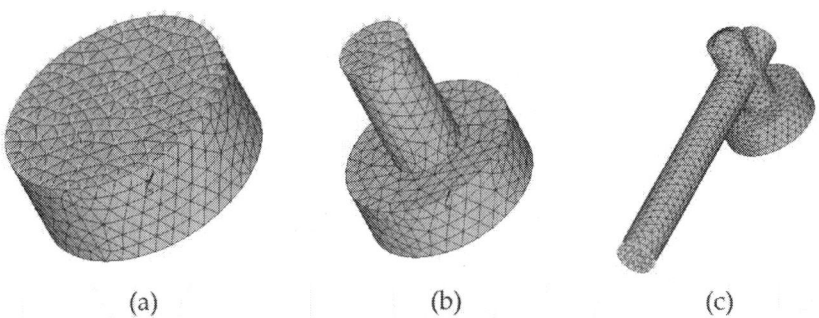

(a)	(b)	(c)

Figure 4. a) Flush adapter mesh; b) Straight adapter mesh; c) Angular adapter mesh.

Experimental Set Up

An experimental unit is conceived to characterize all three adapters, as shown in Fig. 5. The experimental unit is placed into an acoustic reverberant chamber and submitted to an ADF, with a frequency profile shown in Fig. 6, which impinges the plate where the adapters and microphones/conditioners are installed. Care is taken to assure that the plate has structural response similar to that found along the launcher skin. Accelerometers are installed on the plate to measure the acoustically induced structural vibration.

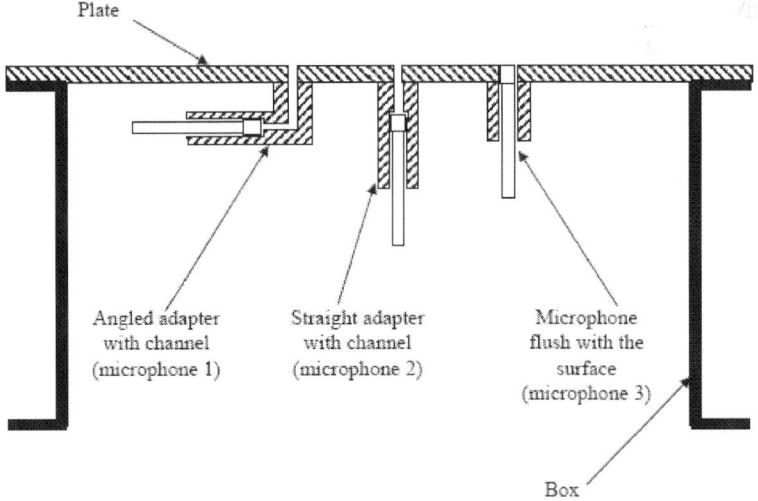

Figure 5. Experimental unit

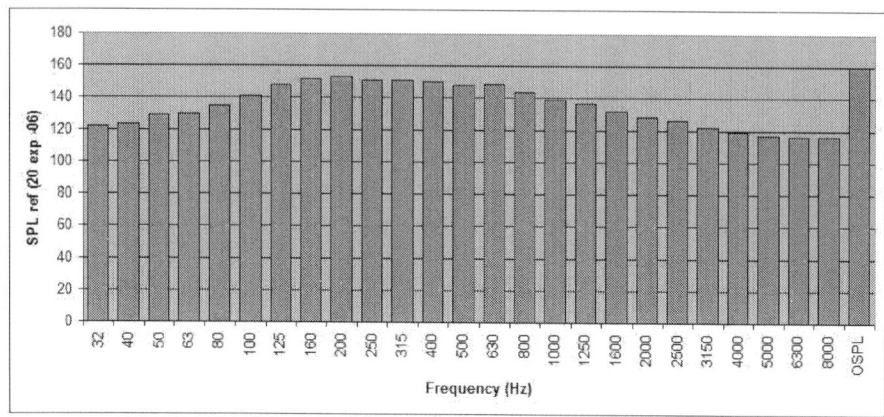

Figure 6. Excitation profile Overall Sound Pressure Level (OSPL) – 160 dB

Methodology for the External Measured SPL Correction

Characterization of the Adapters´ Cavities

The amplitude-frequency characteristics of such adapters must be accounted to determine the acoustic levels measured during the flight mission, since the device channels operate like filters. The SPL measured by the microphone installed with the flush adapter (Fig. 4a) is considered as the reference. Then the adapters transfer functions relating the input signal given by the microphone with the flush adapter and the output signal given by the microphones with the protective adapters (angular or straight) can be obtained by Eqs. (3) and (4)

$$M1(\omega) = H_{M1}(\omega)M3(\omega)$$

(3)

$$M2(\omega) = H_{M2}(\omega)M3(\omega)$$

(4)

where: $H_{M1}(\omega)$ and $H_{M2}(\omega)$: angular and straight adapter transfer functions, respectively. M1, M2, M3: measured SPL into angular, straight and flush adapters.

Results

Analytical and numerical natural frequencies results are compared with those obtained by acoustic testing in Table 2. As pointed out before, the measured SPL with the flush adapter can be read directly. In this way, only the adapters shown in Figures 4 b and 4c are considered.

Table 2. Theoretical x Experimental natural frequencies (in Hz)

Mode	Straight Adapter				Angular Adapter			
	Analytical	FEM	Testing	FEM-Testing Difference (%)	Analytical	FEM	Testing	FEM-Testing Difference (%)
1	5,000	5,396.60	5,202	3.74	2,300	2,204.70	1,953	12.88
2	-	27,032	-	-	-	8,660.30	7,831	10.58

Disagreements between test and predicted resonance frequencies can be explained by possible inaccuracy in microphone installation and complicated shape of the angular channel. The characterization of the adapters' cavities is performed numerically, by calculating H_{M1} (ω) and H_{M2} (ω) (Eqs. (3) and (4)) using FEM. The calculated response functions of the angular and the straight adapter models are compared to those obtained experimentally in Fig. 7 (experimental frequency resolution Δf = 18.78 Hz; numerical frequency resolution Δf = 20.00 Hz).

The theoretical and experimental response functions show good agreement, with a shift in the second resonance peak for the microphone with angular adapter. The errors can be caused by: bad characterization of the ADF spectral distribution into the FEM model; adapter's geometry complexity; microphone installation inaccuracy or a combination of some of these factors.

Equations 3 and 4 show that the external noise (given by M3 (ω)) can be identified by knowing the internal noise and the inverses of the transfer functions for the angular and straight adapters.

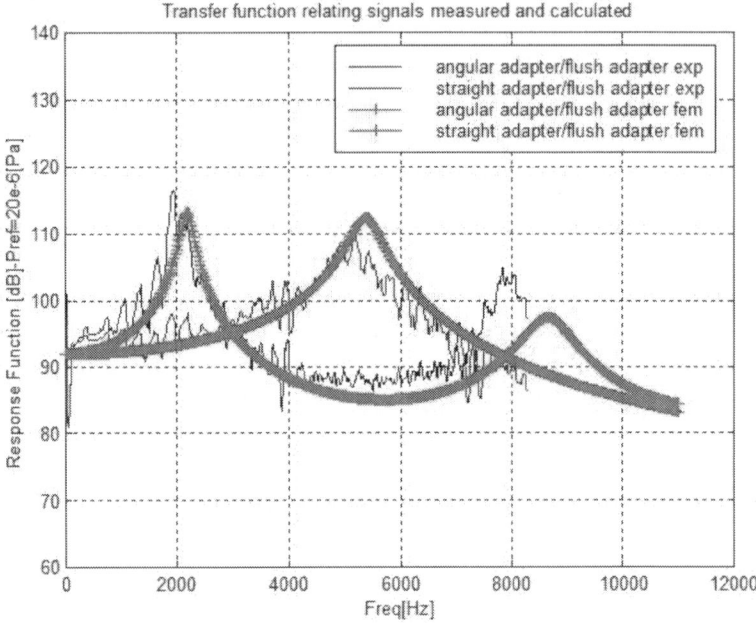

Figure 7. Microphones with adapters transfer functions (numerical and experimental)

VIBRO-ACOUSTIC MODELLING OF PAYLOAD FAIRINGS (PLF)

A complete survey of PLF vibro-acoustic environment must be carried out, in order to determine its inner SPL. In this respect, it is important to have reliable numerical tools that can predict the responses of ELV systems, subjected to in-flight acoustic loads and that enable NCT design.

Low-frequency coupling techniques are used to estimate a PLF dynamic behavior. The fairing body and its inner acoustic domain are analyzed by using Finite Element Method (FEM) and Boundary Element Method (BEM). Structural FEM/fluid FEM and structural FEM/fluid BEM modeling techniques are then applied. In order to simulate the lift off acoustic excitation, an ADF of 145 dB OSPL is applied on the fairing body and coupled calculations are done from 5 to 150 Hz, which yielded the acoustic and skin responses for both models. Modal expansion and semi-modal expansion model techniques are applied, respectively.

For the high-frequency analysis, it is applied the Statistical Energy Analysis (SEA) technique, for a frequency range from 5 to 8,000 Hz. The 145 dB OSPL excitation is applied to the structural panels of the fairing and the acoustic and structural mean responses are calculated.

In view of validating the numerical predictions for the fairing, acoustic test is done to measure the acoustics inside the PLF. The PLF is submitted to 145 dB OSPL in a 1,200 m^3 acoustic reverberant chamber and microphones are positioned in its inner domain.
The implementation of sound absorption blankets is applied as a control technique to attenuate acoustic noise from medium- to high-frequency bands. SEA is a technique for high-frequency analysis; therefore, adequate to assess the influence of blankets on space systems. The generated SEA fluid-structure model is used to calculate internal SPL, with single-, double-, and multi-layered noise control treatments (NCT). Two NCT modeling approaches are used to simulate the effect of blanketing the fairing cavity:

I.   acoustic materials Biot´s parameters, given by the manufacturer;
II.   material samples absorption coefficient, measured in a Kundt Tube.

Model Description
The analyzed fairing is hammerhead type geometry and is composed of the body structure, functional components as electric and pyrotechnic components of the ejection system, mechanisms as well as the exterior cork liner. Figures 8a and 8b show the Brazilian VLS fairing structure.

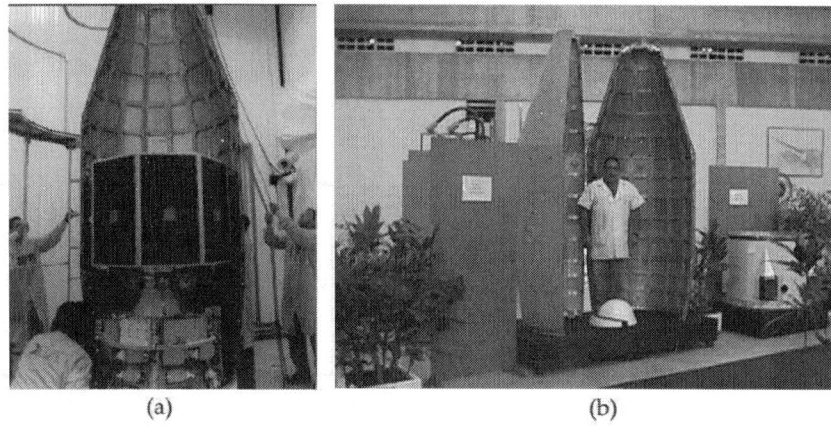

(a)   (b)

Figure 8. a) PLF structure; b) PLF structure

Modelling Methodology

Low-Frequency Modeling (Deterministic) Techniques

In view of predicting the operational fairing cavity SPL, both the dynamic displacements of the fairing structure as well as the acoustic pressure fields at the interior and the exterior side of the fairing should be considered. In this study, however, the fluid-structure coupling interaction between the structural displacements and the exterior acoustic pressure field is neglected. The exterior acoustic pressure is assumed to be a known external excitation for the vibro-acoustic system, consisting of the fairing body and the internal acoustic cavity. The FEM and BEM are the most appropriate numerical techniques for the (low-frequency) dynamic analysis of this type of vibro-acoustic system.

FEM based models for coupled vibro-acoustic problems are most commonly described in an Eulerian formulation, in which the fluid is described by a single scalar function, usually the acoustic pressure, while the structural components are described by a displacement vector. The resulting combined FEM/FEM model in the unknown structural displacements and acoustic pressures at the nodes of, respectively, the structural and the acoustic FEM meshes are [20],

$$\left(\begin{bmatrix} K_S & K_C \\ 0 & K_A \end{bmatrix} + j\omega \begin{bmatrix} C_S & 0 \\ 0 & C_A \end{bmatrix} - \omega^2 \begin{bmatrix} M_S & 0 \\ -\rho K_C^T & M_A \end{bmatrix} \right) \begin{Bmatrix} w_i \\ p_i \end{Bmatrix} = \begin{Bmatrix} F_{Si} \\ F_{Ai} \end{Bmatrix}$$

(5)

Where:

KS – Structural stiffness matrix MS – Structural mass matrix CS – Structural damping matrix

KC – Fluid – structure coupling matrix

MA – Fluid mass matrix KA – Fluid stiffness matrix CA – Fluid damping matrix

wi – Structural displacements nodal vector pi – Acoustic pressures nodal vector

FSi – Force nodal vector FAi – Acoustic sources nodal vector

In comparison with a purely structural or purely acoustic FEM model, the coupled stiffness and mass matrices (Eq. (5)) are no longer symmetrical due to the fact that the force loading of the fluid on the structure is proportional to the pressure, resulting in a cross-coupling term KC in the coupled stiffness matrix, while the force loading of the structure on the fluid is proportional to the acceleration, resulting in a cross-coupling term $M_C = -\rho K_C^T$ in the coupled mass matrix.

Low-frequency vibro-acoustic problems can also be modeled by describing the structural behavior in a FEM model and the fluid behavior in a BEM model. In the same way as in the FEM/FEM technique, deterministic FEM/BEM models are usually described by acoustic double layer potential and structural displacement, which are the field variables. Equation (6) presents the resulting combined FEM structural displacements and BEM acoustic pressure differences at the nodes, for a coupled FEM/BEM mesh [23].

$$\begin{pmatrix} K_S + j\omega C_S - \omega^2 M_S & L_C \\ L_C & \dfrac{D}{\rho_0 \omega^2} \end{pmatrix} \begin{Bmatrix} w_i \\ \mu_i \end{Bmatrix} = \begin{Bmatrix} F_{Si} \\ F_{Ai} \end{Bmatrix}$$

(6)

Where LC is the fluid-structure coupling matrix, D is the BEM acoustic matrix of coefficients and μi is the nodal vector of double layer potentials. In deterministic models, the dynamic variables within each element are expressed in terms of nodal shape functions, usually based on low-order (polynomial) functions. Since these low-order shape functions can only represent a restricted spatial variation, a large number of elements is needed to accurately represent the oscillatory wave nature of the dynamic response. A general rule of thumb states that for fluid-structure interactions, at least 6 (linear) elements per wavelength are required to get reasonable accuracy. Since wavelengths decrease for increasing frequency, the FEM model sizes, computational efforts and memory requirements increase also with frequency. As a result, the use of FEM and BEM models is practically restricted to low-frequency applications. In comparison with uncoupled structural or acoustic problems, this practical frequency threshold becomes significantly smaller for coupled vibro-acoustic problems, since a structural and an acoustic problem must be solved simultaneously. Moreover, the matrices in a coupled model are no longer symmetrical, so that less efficient non-symmetrical solvers must be used. As a consequence, the computational effort, involved with the use of coupled FEM/FEM and FEM/BEM models for real-life vibro-acoustic engineering problems, becomes large at very low frequencies.

In order to obtain coupled vibro-acoustic response predictions within reasonable computational efforts, the dimensions of the FEM/FEM problem (Eq. (5)) have to be reduced. The most applied technique for model reduction is the modal superposition technique, which expresses the unknowns of the system in terms of a modal basis, resulting in a set of unknown modal participation factors, whose size is much smaller than the size of the original set of unknowns. A modal expansion in terms of uncoupled structural and acoustic modes is performed by using computationally efficient symmetric eigenvalue algorithms and requires much less computational effort than the use of vibro-acoustic (coupled) modes. However, a large number of high-order uncoupled acoustic modes is required to accurately represent the normal displacement continuity along the fluid-structure interface.

In a FEM/FEM virtual prototype, a modal expansion in terms of uncoupled structural and uncoupled acoustic modal bases is used, in order to keep the computational efforts within reasonable limits. On one hand, structural wavelengths are much smaller than acoustic wavelengths, so that the structural FEM mesh of the fairing must be finer than the acoustic FEM mesh of the inner cavity. On the other hand, due to the continuity of the normal structural and fluid displacements along the fluid-structure coupling interface, both meshes must be compatible in this region. In this framework, the following modeling methodology is adopted: 1) A fine FEM mesh of the fairing is used for the construction of the uncoupled structural modal data basis. 2) The resulting modes are then projected onto a FEM coarse mesh of the fairing structure. 3) For the acoustic cavity FEM mesh, the same mesh density is used along the fluid-structure coupling interface as the PLF structural coarse mesh, while the mesh density is slightly decreased towards the central axis of the cavity. 4) The uncoupled modes, resulting from this acoustic FEM mesh, together with the projected structural modal basis are used in a coupled FEM/FEM model. It is important to highlight that the coarse structural mesh has only the shells of the fairing structure, while all reinforcing beams are omitted, since it is assumed that these stiffeners have no significant effect on the fluid-structure coupling interaction, while their presence would increase the computational load of the modeling process.

For the case of the FEM/BEM problem (Eq. (6)), the modal expansion cannot be used, since the frequency dependency of the matrix coefficients in the acoustic part prohibits a standard eigenvalue calculation. Such as, the semi-modal approach, which uses only the expansion of the structural modal data basis, is applied. As mentioned above, BEM drawbacks as fully population of the matrices, complex and frequency dependent models result in a coupled FEM/BEM model less efficient than coupled FEM/FEM model. Therefore, the rule of thumb of 6 (linear) elements per wavelength becomes prohibitive for the actual fairing fluid-structure study. Such a way, a coarsest structural mesh may be generated and the same adopted frequency range for FEM/FEM model is kept for this FEM/BEM model, even considering that the structure has not enough discrete density. However, the modal data basis calculated using the structural fine mesh can assure good results for the structural displacements, in this fluid-structure model, since the expansion in terms of such a data basis is used on these Frequency Response Analysis (FRA) computations. The displacement continuity of the structural and acoustic meshes (same density in the fluid-structure interface) is considered to perform the link and calculate the coupled dynamic skin displacements and acoustic cavity pressure responses, for both FEM/FEM and FEM/BEM models.

High-Frequency Modeling (Statistical) Technique

A characteristic of high-frequency analysis is the uncertainty in modal parameters. The resonances and mode shapes show great sensitivity to small variations of geometry, construction and material properties. In addition, programs used to evaluate mode shapes and frequencies are known to be inaccurate for higher modes. In light of these uncertainties, a statistical model of the dynamic parameters seems natural and appropriate. As an alternative method for higher frequency analysis of the inner cavity of fairings, Statistical Energy Analysis (SEA) approach is proposed. This approach is the description of the dynamic system as a member of a statistical population or ensemble, whether or not the temporal behavior is random. SEA emphasizes the aspects of this field dynamical study.

The SEA equations express the energy balance of different subsystems in a model [24]. Some subsystems have direct power input of an independent source, e. g. an excitation force on a structural component, a sound power source in an acoustic medium etc. In general, subsystems can receive power (input power from external sources), dissipate power (internal losses due to damping) and exchange power with other subsystems to which they are coupled (losses due to coupling). SEA fundamental hypothesis as dissipation losses in relation to the energy variable and modal energy proportionality from connected subsystems are used to yield the SEA matrix equation of complex structures. The distribution of the dynamical response in the system due to some excitation is obtained from the distribution of the energy among the mode groups, based on a set of power balance equations for the mode groups.

Fem Structural Meshes

The fairing body is divided in five surfaces. The surfaces are discretized by using 4-noded quadrilateral shell elements, while 2-noded beam elements are used for the circumferential and the axial stiffeners. To account for the mass of the cork blanket on the exterior fairing surface, a distribution of concentrated mass elements are attached to the fairing nodes.

A total of 174 structural modes in a frequency range up to 220 Hz have been identified. Table 3describes the main structural modes calculated by using FEM in the range up to 150 Hz.

Table 3. Fairing body structural modes

mode	frequency (Hz)
first bending	38.637
first breathing	77.821
second breathing	92.694
first longitudinal	108.749
first torsion	125.772
second bending	150.735

Table 3 shows that the first structural bending mode of the fairing is identified at 38.6 Hz, while the second mode is at 150.7 Hz. Figures 9 a and 9b show the referred structural modes.

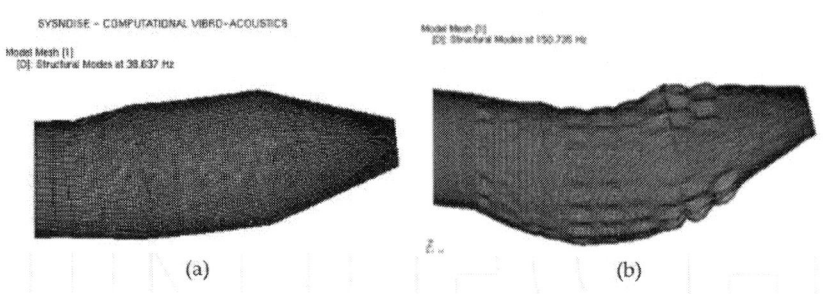

(a)   (b)

Figure 9. a) First structural bending mode; b) Second structural bending mode

Fem and Bem Acoustic Meshes

The acoustic FEM mesh consists of 119,577 nodes and 110,238 elements (106,050 8-noded hexahedral elements and 4,188 6-noded pentahedral elements). The cavity is considered filled with air at 15° C. The cavity's bottom and top faces are assumed to be acoustically closed (rigid walls). The acoustic mesh generation takes into account the meshes compatibility on the fluid-structure interface.

A total of 80 acoustic modes in a frequency range up to 566 Hz were identified. Acoustic wavelengths are bigger than structural wavelengths. Such that, a large number of high-order uncoupled acoustic modes is required to accurately represent the normal displacement continuity along the fluid-structure interface. That is why higher frequency range is used to describe the acoustic modal behavior of the fairing. Table 4 describes the acoustic modes in the frequency range up to 150 Hz.

Table 4. Fairing cavity acoustic modes

Mode	Frequency (Hz)
rigid body	0
first longitudinal	63.491
second longitudinal	112.129

It can be noticed in table 4 that the first and second acoustic modes of the fairing cavity are identified at 63.5 Hz and 112.1 Hz, respectively. Figures 10a and 10b show the referred acoustic modes.

The BEM acoustic mesh is a 2-D coarsest mesh. Therefore, as the coupled FEM/BEM equation is frequency dependent (Eq. (6)), the acoustic modes are not considered in the acoustic pressure calculations (semi-modal reduction model).

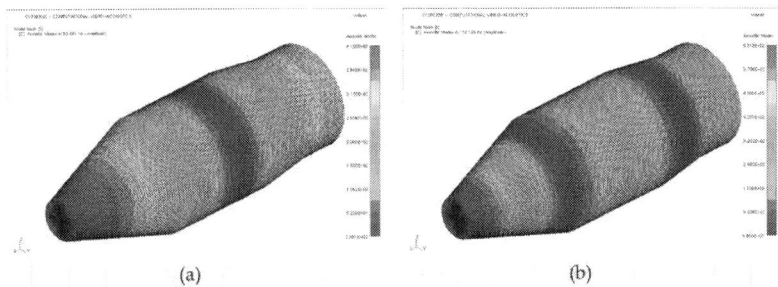

(a)	(b)

Figure 10. a) First acoustic longitudinal mode; b) Second acoustic longitudinal mode

Model Excitation

A uniform pressure loading is simulated by applying a normal point force varying harmonically on all nodes of the fairing shell elements. The force value is defined such that the total load is equivalent to a uniform pressure of 145 dB OSPL. Link of the acoustic and structural parts is done as well as the structural modal data basis is projected to the coarse and coarsest meshes.

In this way, all the meshes, modal data bases and excitation, needed to perform low-frequency calculations, using coupling fluid-structure techniques are ready. Next step is to perform FRA calculations for both models.

Sea Fairing Vibro-Acoustic Model

The fairing body is divided in four surfaces, as shown in Fig. 11a. To account for the rib-stiffened plates of the surfaces 2, 3 and 4, the SEA structural fairing model considers connected plates and beams (longitudinal and circular). The plate structural subsystems are generated as singly curved shells and uniform plates. Shell surface 1 has a thickness of 3mm and is modeled as a simple plate of aluminum (E=72 GPa, v=0.29, ρ=2750 kg/m^3), while the other three surfaces are 0.8 mm thick and made of an aluminum alloy (E=72 GPa, v=0.29, ρ=7000 kg/m^3). The circular and longitudinal beams are modeled by assigning the same material as the shells of the surfaces 2, 3 and 4 (Figure 11a). Damping loss factors of 1% (for flexure, extension and shear propagating waves) are assigned to the plates and beams subsystems, in order to account for the internal loss factors.

A total of 72 beams (44 longitudinal and 28 circular) and 8 shells (02 singly curved shells of the adaptor, 02 singly curved shells of the lower cone, 02 singly curved shells of the main cylinder and 02 singly curved shells of the upper cone) compose the structural SEA model. Figure 11a shows the SEA plates and beams generated to model the VLS-1 fairing structure. The external blanketed treatment of cork on the surfaces 2, 3 and 4, was simulated in this model as material addition. The layered area and the density of the cork were considered to assign this mass.

The acoustic environment inside the fairing was generated by starting from the structural model. This acoustic cavity was created considering air at 15° C as the fluid as well as the dimensional parameters of the fairing. The top and bottom face of the cavity were assumed to be acoustically closed. Figure 11b presents the 3D acoustic cavity of the fairing.

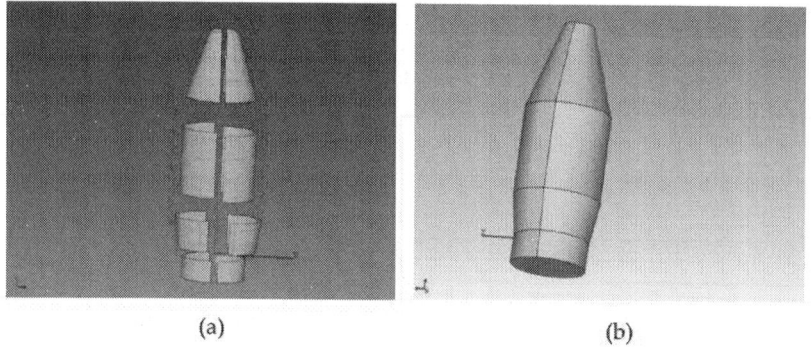

(a) (b)

Figure 11. a) Structural subsystems (shells, circular beams and longitudinal beams) b) Acoustic cavity of the fairing

The coupling boundary between all the structural and acoustic subsystems is modeled to consider the transmission of power across the junctions. A junction is comprised of connections to any number of coincident subsystems. As a result, all the subsystems that share common nodes are connected by point, line and area junctions and all the appropriate wavefields are connected as well as the corresponding coupling loss factors (CLF) between subsystems are created.

The estimated sound pressure levels at the lift off are assigned to the SEA model. Only elements with large surface areas, as plates and panels, are considered to be susceptible to acoustic excitation ([11], [24] and [25]).

An ADF of 145 dB OSPL (Fig. 12) is applied to the plates of the SEA fairing model, which simulates the power input into a structural plate or shell element.

Analysis Results

In view of having a complete knowledge of the fairing dynamic vibro-acoustic behavior, the fairing structural skin as well as its inner acoustic domain responses should be presented. However, since this chapter concerns acoustics, the body structural displacements are not presented here. Below, the obtained results of the acoustic behavior applying vibro-acoustic low-frequency and high-frequency analysis techniques predictions are presented.

Low-Frequency Techniques

FEM/FEM response calculations

A modal expansion in terms of 174 uncoupled structural and 80 uncoupled acoustic modes is used for the coupled calculations. A modal damping of 1% is assigned to all structural modes. All calculations are performed with a frequency resolution of 1 Hz. Figure 13 shows the low-frequency acoustic pressure spectra of the PLF for the case of a uniform exterior pressure loading, using FEM/FEM coupling analysis. It can be seen that the low-frequency pressure is dominated by the first longitudinal mode around 63.5 Hz and the second longitudinal mode around 112.1 Hz.

FEM/BEM Response Calculations

The same structural modal expansion as used for FEM/FEM is used for this FEM/BEM response calculations. Due to the frequency dependency of the boundary integral equation, the acoustic modal basis can not be used. A damping of 1% is assigned to all structural modes. All calculations are performed with a frequency resolution of 2 Hz. Figure 13 presents a comparison of the computed inner cavity space averaged acoustic pressure using FEM/FEM and FEM/BEM techniques.

High-Frequency Technique

The energy balance (levels and interactions) between different subsystems of the SEA model is calculated. The interest frequency range is 5 to 8,000 Hz, by third octave bandwidth. As mentioned before, SEA technique is more effective in higher frequencies, where dynamic systems present higher modal density. The vibro-acoustic responses of the fairing, using SEA technique, are shown in Fig 13.

It is important to highlight that for the low-frequency range, SEA analysis results are not reliable, since the accuracy of the SEA technique is proportional to the modal density [24].

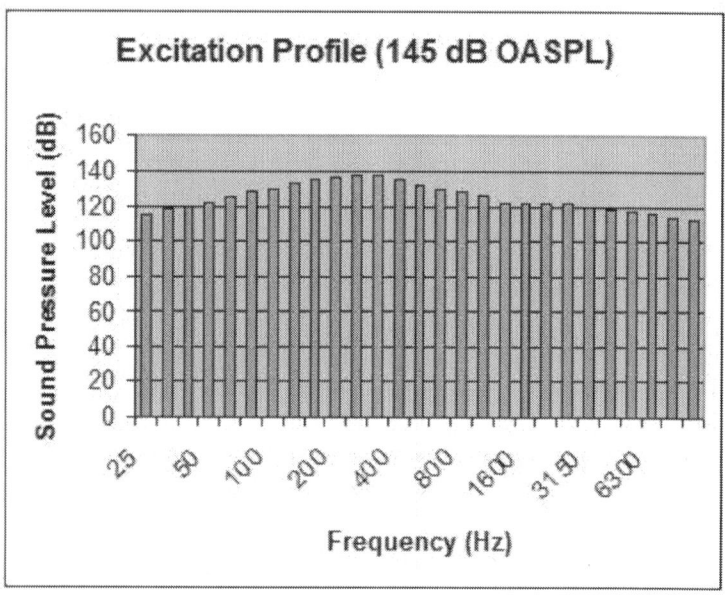

Figure 12. Acoustic diffuse field at lift off

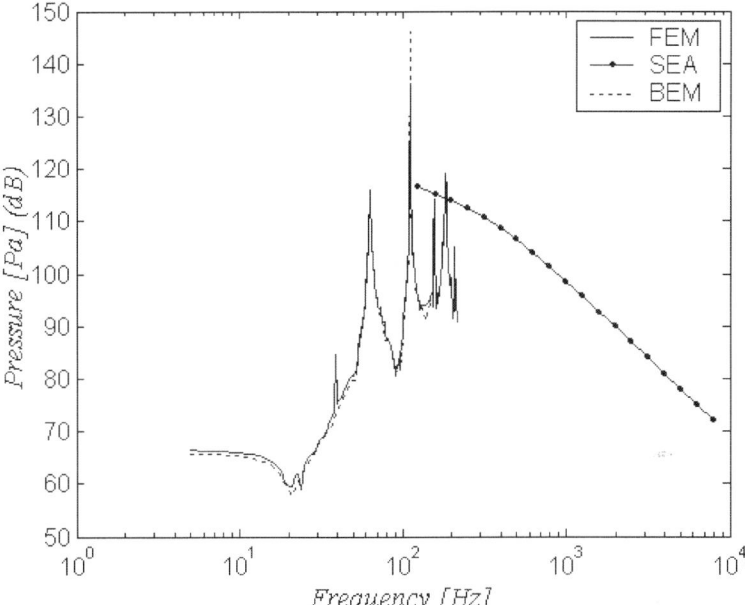

Figure 13. Acoustic Response Inside PLF

Considering the accuracy, advantages and drawbacks of the deterministic and statistical techniques, each of them is successfully applied in different frequency ranges. Such that, for the analyzed PLF, valid response results using deterministic techniques are assumed up to 150 Hz, while valid SEA results are assumed from 300 up 8,000 Hz. It is important to mention that in the "twilight zone" or medium frequency bandwidth (from 150 to 300 Hz), where deterministic models are inaccurate and present prohibitive computation time for the calculations and where the high modal density requirement is not yet accomplished for SEA, both results may be considered, as shows figure 13.

Model Validation

The fairing structure was positioned inside an acoustic chamber and excited with an ADF of 145 dB OSPL. Eight control microphones were positioned inside the reverberant chamber, which feedback the control system. Four measurement microphones were located in the acoustic cavity of the PLF. The measured space averaged SPL is compared with the theoretical acoustic responses, computed using the virtual prototypes (FEM and SEA models) (Fig. 14).

The calculated internal acoustic frequency response function shown in figure 13 may be transformed into 1/3 octave band responses to be compared with the experimental (measured) results. Figure 14presents the 1/3 octave comparisons for the frequency bandwidths ranging from 31.5 up to 8,000 Hz. It can be noticed, that experimental and calculated low-frequency responses have good agreement, presenting more significant differences only on the 1/3 octave bands 31.5, 40 and 50 Hz. This is because the low-frequency modes of the acoustic chamber are not well excited. However, in the regions where the cavity response is dominant (63 Hz and 112 Hz), differences are pretty small.

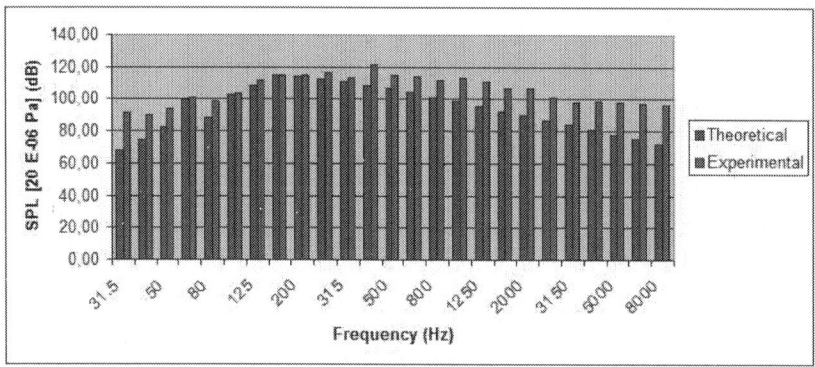

Figure 14. High-frequency theoretical X experimental comparison

For the higher frequencies, a more reasonable comparison should be done using Power Spectral Density (PSD) [28], since SEA calculations may be interpreted as mean values of energetic response functions when averaged at a given frequency over an ensemble of similar systems, differently of peak values resulting from deterministic approaches. However, a qualitative comparison can be presented for 1/3 octave bands from 160 up to 8,000 Hz, since one keeps in mind that mean values and the predicted magnitudes yielded by SEA should under estimate the dynamic response with a certain (acceptable) variance. Valid SEA results are assumed from 300 up 8,000 Hz, since the minimum five modes by bandwidth (modal density) requirement becomes true starting from 300 Hz.

At the beginning phases of space projects, the assessment of the effect of using different passive techniques for acoustic environment alleviation to be applied to PLF is an important issue. One of the main applications of numerical control prediction is the decision, still in the early product development phase, which design version is the most appropriate from the noise control point of view. By introducing the concept of sensitivity analysis, product development can be performed in a more systematic way. In order to predict the efficiency of a NCT, one compares the effects of design modifications.

In this framework, different blanket layers are implemented on the PLF elasto-acoustic virtual prototype and the effects of these NCT implementations are assessed. Since blankets acoustic absorption depend on certain material parameters, two blanket modeling approaches are assessed as follows: material physical Biot's parameters as density, porosity, resistivity, tortuosity, viscous and thermal characteristic lengths, given by the blanket manufacturer and measured normal incidence absorption coefficients of material samples.

Furthermore, the influence of the NCT thicknesses and the presence of air-gaps between blankets are analyzed. Biot's parameters and absorption coefficient approaches are implemented in the coupled elasto-acoustic SEA model of the PLF.

For the Biot's parameters approach, an explicit model of the inserted material is considered, based on the physical properties of individual layers, which are accounted in the SEA model. Six types of glass wools are analyzed and the SPL inside the fairing are calculated. The wools' densities are given in pounds per cubic feet (pcf - 1 lb/ft^3 = 16.02 Kg/m^3). A thickness of 7,62 cm is adopted for almost all glass wools, but the two 1.2pcf glass wools, that presents particular behavior, which adopted thicknesses were 0.19 cm and 0.38 cm. The best performing material is chosen and a comparison between different thicknesses and percentages of layered surfaces of the fairing is done, considering the final weight of the applied NCT. The materials used were glass wools described in Table 5. The wools' Biot's parameters can be found in [4].

Table 5. Glass wools used

Material characteristcs						
Density (pcf)	0.34	0.42	0.6	1.2	1.2	1.5
Thickness (inches)	3	3	3	0.75	1.5	3

On the other side, the measured absorption coefficient of multi- and single-layered samples of glass wools of 0.42 and 1.0 pcf were considered. According to [26], air gaps between materials increase the acoustic absorption at low-frequencies. For this case, samples with two different air gaps are positioned into a Kundt tube. The single-layered samples are 3.50 cm thick, while combinations are done with samples of 1.75 cm thick. Other configurations were assembled with air gaps of 1.0 and 3.0 cm between samples. Figure 15 shows the sample combinations. All the measured absorption coefficients are shown in Fig. 16. These absorption coefficients are assigned on the fairing vibro-acoustic model and SPL are calculated.

The PLF acoustic responses for different NCT configurations are shown in Fig. 17. Notice that the insertion of the 0.34pcf glass wool − 7.62 cm yields almost 20 dB of attenuation (chosen as the best performing material). The assessment of the thickness influence is done by assigning 0.34pcf glass wools of 7.62, 10.16 and 12.7 cm thicknesses, with total NCT weights of 3.90, 5.30 and 6.60 Kg, respectively. Figure 18 shows the internal SPL one-third octave distribution, as well as the OSPL.

Figure 19 shows SPL and OSPL from 50 to 8,000 Hz, for the NCT described in Fig. 16, without air gaps. A 3.50 cm double-layered blanket (0.42pcf/1.0pcf) is compared with two single-layered NCT. Notice in this figure that NCT decrease the internal OSPL from 132 dB to 128 dB. Figure 20 shows that single-layered treatment with 1.0 pcf and air gaps presented better results. One can see the air gap effect, since the SPL close to 100, 315 and 500 Hz are higher mainly when the 1.0 pcf material with 3.0 cm air gap is applied. The calculations yielded 127.5 dB OSPL inside the fairing cavity. This means that a gain of approximately 3.0 dB at 100 Hz bandwidth can be obtained, yielding an overall gain of 1.0 dB, approximately. However, air gap installation can be limited, due to fairing internal space. In this case, it is preferable to install the blanketed treatment distant from the panels by small air gaps, instead of bonded, once this installation configuration presents higher transmission loss [10].

Figure 15. Double-layered 0.42pcf (yellow)/1.0pcf (orange), 1.75 cm thick each.

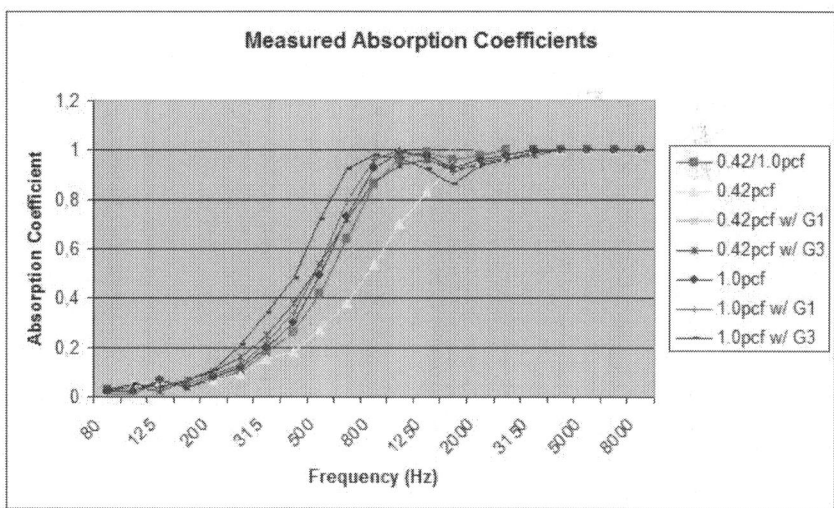

Figure 16. Measured absorption coefficients*G1 and G3: air gaps of 1.0 cm and 3.0 cm

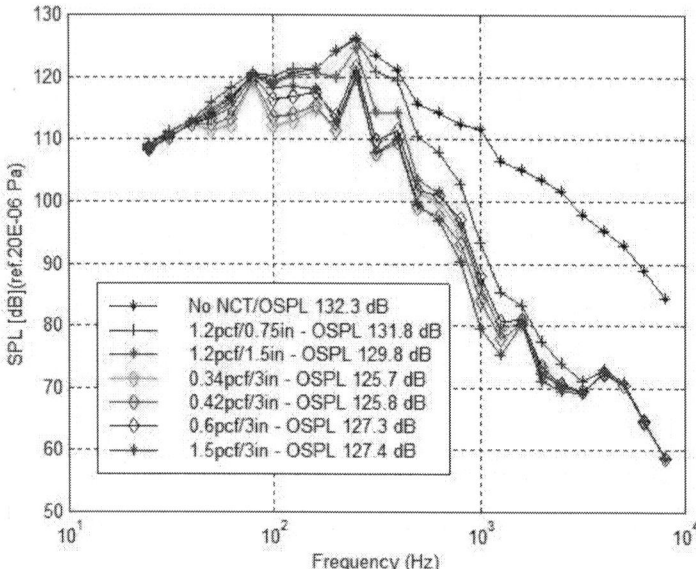

Figure 17. SPL for different blankets

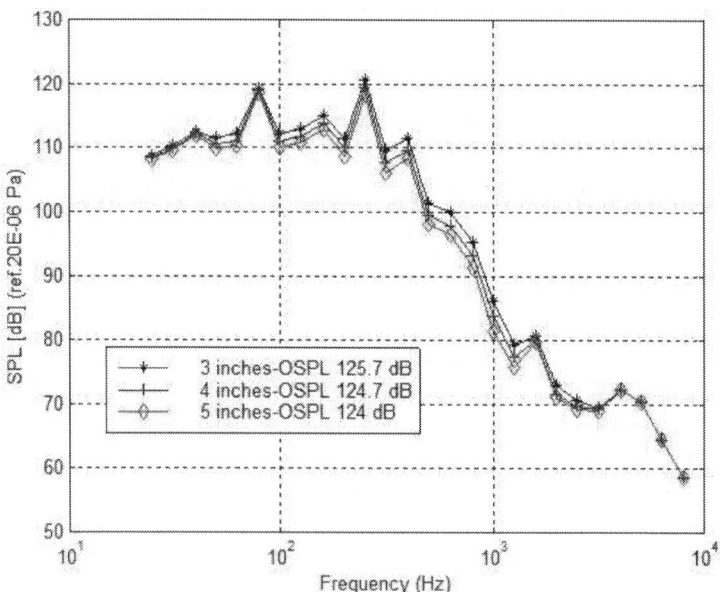

Figure 18. SPL for different thicknesses

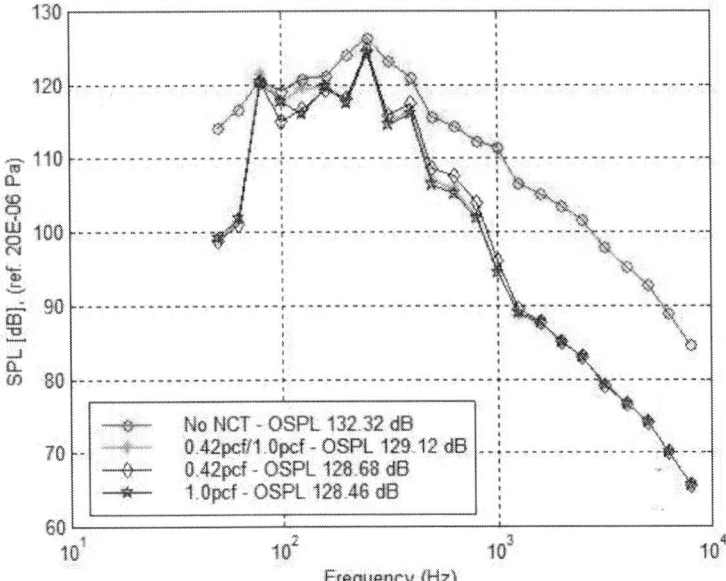

Figure 19. SPL for NCT without air gap.

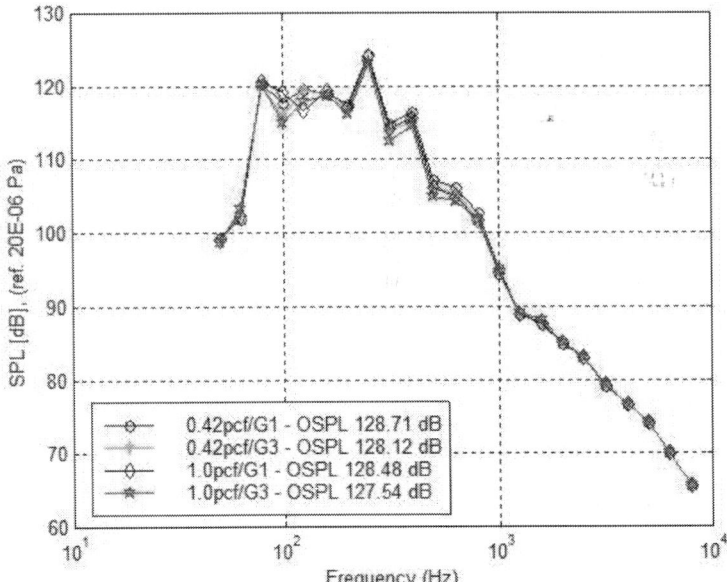

Figure 20. SPL for NCT with air gaps (*G1 and G3: air gaps of 1.0 cm and 3.0 cm)

COMBUSTION INSTABILITIES OF LIQUID PROPELLED ROCKET ENGINES DUE TO CHAMBER COMBUSTION ACOUSTICS

Combustion instabilies have been present in the development of LPRE over the last decades. There are basically three types of combustion instabilities: low-frequency (LF), medium-frequency (MF) and high-frequency (HF). LF instabilities, also called chugging, are caused by pressure interactions between the propellant feed system and the combustion chamber. MF instabilities, also called buzzing, are due to coupling between the combustion process and the propellant feed system flow. The HF instabilities are the most potentially dangerous and not well-understood ones. It occurs due to coupling of the combustion process and the chamber acoustics [27].

The presence of acoustic combustion instabilities must be considered still in development phase, although combustion instabilities can be clearly identified only during firing tests. In [13], it was described that instability can be verified when the power spectrum of the acoustic pressure levels, measured during burning tests, is analyzed. When an oscillation is observed, i.e., combustion instability, sound pressure peaks with well-defined magnitudes summed to the background noise are present. These peaks are correlated to the resonance frequencies of the combustion chamber. This phenomenon can cause instabilities and consequent unexpected behavior such as efficiency loss or even explosion of the engine. In this framework, the engine acoustic cavity characterization becomes an important issue to be investigated.

Acoustic behaviour of chambers is usually determined by doing cold tests measurements (without combustion). Acoustic dynamics in combustion environments are obtained by shifting the cold test resonant frequencies by a scalar factor defined by the ratio of sound velocity at the cold test temperature and at real operation temperature [15].

In view of attenuating acoustic pressure oscillations inside combustion chambers, reactive techniques as Helmholtz Resonators (HR), among others, are widely used ([13] and [16]). These devices are specially designed to attenuate oscillations at discrete resonance frequencies (pure tones). HR have been applied as combustion stabilization devices for solid motors and liquid rocket engines, with success. It could be noted in literature that they are used in a set of dozens or even hundreds in each chamber cavity, distributed along the walls or in a single row along the injector periphery [28].

This item describes a procedure for cold test acoustic characterization of LPRE combustion chambers. Firstly, the acoustical dynamic characterization of a combustion chamber is done and a typical longitudinal resonant frequency is chosen to be attenuated. A HR is designed (tuned at the chosen frequency) and applied to the mock up face plate. A LPRE mock-up [14] was used as experimental model. This test rig faithfully represents the internal acoustic cavity of the original engine. This procedure is followed by doing virtual prototypes of the combustion chamber. The acoustic natural frequencies and mode shapes are numerically calculated by a FEM model and validated through acoustic experimental modal analysis [29].

Experimental Acoustic Modal Analysis (Eama)

Experimental Modal Analysis (EMA) is a well-applied technique in structure dynamics. However, due to the development of commercial acoustic sources, EAMA can be a suitable choice in view of extracting the acoustical Frequency Response Functions (FRF). In addition, the mathematical approach of structures modal parameters extraction can be applied to acoustic systems. [29].

In order to check the mutual orthogonality among modes from a modal model and to compare modes between different modal models (i.e., experimental and numerical solutions), the Modal Assurance Criterion (MAC) was used. This criterion indicates the degree of linear dependence between two eigenvectors and can be described as Eq. (7) [30].

$$MAC_{ijm} = 100 \cdot \frac{\left| \{\phi_{im}\}^T \{\phi_{jm}^*\} \right|^2}{\{\phi_{irm}\}^T \{\phi_{im}^*\}\{\phi_{jm}\}^T \{\phi_{jm}^*\}}$$

(7)

where: indexes *i* and *j* denotes modes obtained by different methods.

Helmholtz Resonator

Helmholtz Resonators are widely applied in order to suppress or attenuate the acoustic pressure inside cavities, rooms and other volumes. A HR consists of a small volume connected to a bigger cavity (the combustion chamber, in this case) through an orifice by a flanged neck. The dimensions of the HR must be much smaller than the acoustic wavelength of interest, in order to consider the resonator as lumped elements coupled to a geometric discontinuity. The coupling condition is that the oscillatory volume flow in the neck is equal to that imposed on the fluid inside the cavity, neglecting the elastic property of the air in the neck [30].

A typical HR is shown in Fig.21 (left), being d the neck diameter, D the cavity diameter, Vc the volume cavity, l the neck length and L the cavity length. $P1$ is the incident acoustic pressure and $P2$ is the cavity pressure. The gas motion in the HR coupled in an acoustic cavity can behave equivalently to a mass-spring-dashpot system (Fig.21, centre). The system can be divided into three distinct elements. The fluid enclosed in the neck behaves as an uncompressible gas, and its mass correspond to the m element of the mechanical system. The air inside the cavity is compressible and stores potential energy, representing the mechanical stiffness k. The mechanical damping element (c) is represented by two factors: (i) the open-end of the neck radiates sound, introducing a radiation resistance and (ii) the gas movement in the neck introduces a viscous resistance. Considering the electrical analogue (Fig.21 right), the acoustic compliance C (analogous to electrical capacitance) is related to the stiffness of the air in the cavity, the acoustic inertance M (analogous to electrical inductance) is associated to the inertia element (mass) and the acoustic resistance R (analogous to electrical resistance) is related to the dissipative components stated above.

Considering that the gas beyond the end of the neck moves as a whole with the gas inside the neck, it is necessary to use an effective length $leff$ which is bigger than the true length l of the neck [19]. The effective length $leff$ is obtained by adding a mass end correction δ, which is empirically determined. In [28], it was presented a complete set of recommended equations for mass end correction, depending on the adopted considerations. For the purpose of this work, the appropriated equation is defined by:

$$\delta = 0.85d\left(1 - 0.7\sqrt{AR}\right) \text{ for } AR < 0.16$$

(8)

where AR is the Area Ratio (An / Ac), being An and Ac the neck cross-sectional area and the cavity cross-sectional area, respectively. The effective length is calculated as $leff = l + \delta$.

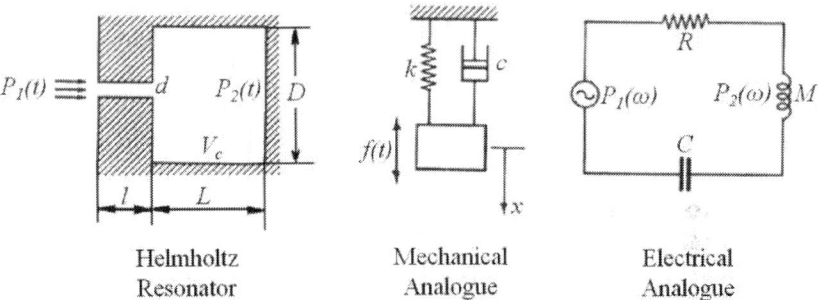

Helmholtz Resonator Mechanical Analogue Electrical Analogue

Figure 21. Helmholtz Resonators scheme and its analogues

The definition of acoustic inertance (M) applied to the Helmholtz resonator gives:

$$M = \frac{m}{A_n^2} = \frac{\rho l_{eff} A_n}{A_n^2} = \frac{\rho l_{eff}}{A_n}$$

(9)

where ρ is the air density and m is the effective mass.

The acoustic compliance C is defined as the volume displacement that is produced by the application of unit pressure [19]. By applying this definition to HR, one obtains:

$$C = \frac{V_c}{\rho c^2}$$

(10)

where c is the velocity of sound.

The acoustic resistance in the neck (R) was approximated as the dissipation associated with viscous forces, considering the dynamic viscosity μ [28]:

$$R = \frac{8\pi\mu l}{A_n}$$

(11)

The acoustic impedance Z of the HR is:

$$Z = R + j\left(\omega M - \frac{1}{\omega C}\right)$$

(12)

As can be seen, the acoustic impedance is determined by the geometric and mechanical properties of the resonator. The resonance will occur when the acoustic reactance equals zero:

$$\omega M - \frac{1}{\omega C} = 0$$

(13)

The resonance frequency can be determined by considering that the dimensions of the resonator are much smaller than the wavelength of interest:

$$\omega_0 = \sqrt{\frac{1}{MC}} = c\sqrt{\frac{A_n}{l_{eff} V_c}}$$

(14)

The resonance's sharpness of a HR can be quantified by its quality factor Q, given by:

$$Q = \frac{\omega_0 M}{R}$$

(15)

Finite Element Model

The cavity of a LPRE combustion chamber was analyzed using FEM in configurations without and with resonators. The first was modeled using 11,136 linear solid hexahedral elements, 12,510 nodes (12,093 degrees of freedom) and the second was modeled using 38,052 linear solid tetrahedral elements, 7,493 nodes (7,399 degrees of freedom). Both meshes are shown in Fig. 22. The fluid is air at 15° C. The eigenfrequencies were calculated from 0 to 2,400 Hz. Nodal pressures on the openings were assigned to zero.

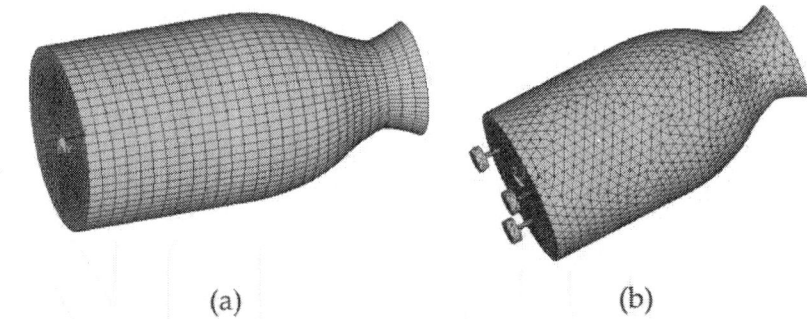

(a)   (b)

Figure 22. a) Cavity without resonators; b) Cavity with resonators

Experimental Setup

Figure 23 shows the experimental setup. The mid-high frequency volume acceleration source is composed by a driver, a tube and a nozzle, where it is installed a volume acceleration sensor. This source produces a voltage signal proportional to the volume acceleration [m^3/s^2] variation, with a nominal frequency range of 200 up to 8,000 Hz. This source nozzle was installed in the mock-up plane surface as shown in Fig. 23.

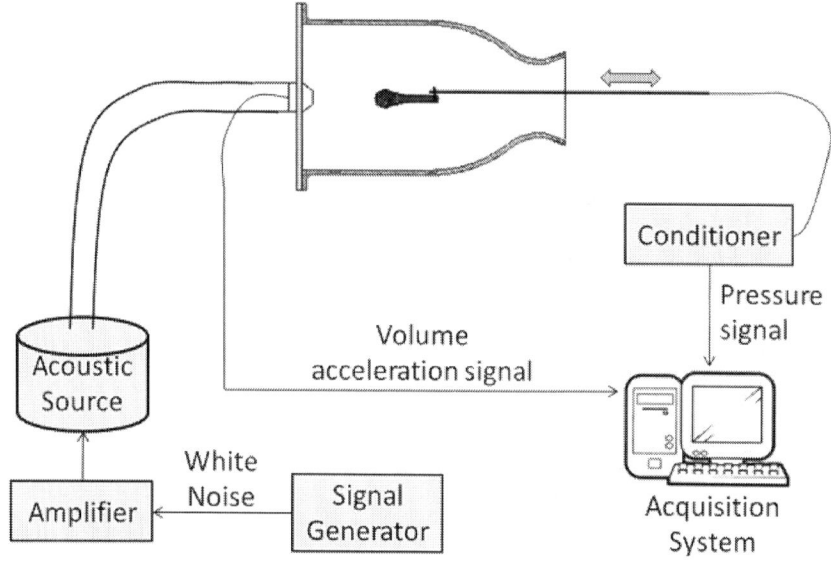

Figure 23. Measurement setup

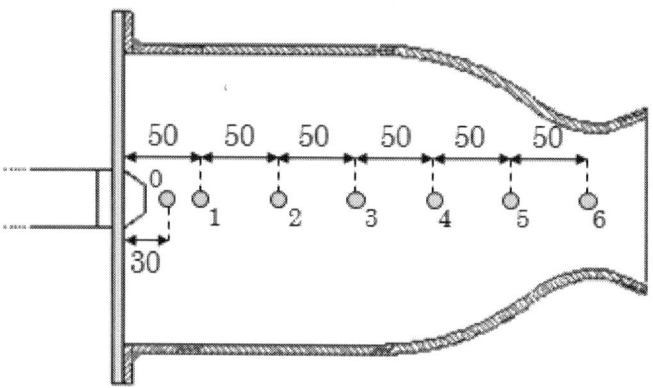

Figure 24. Source and microphone positions (distances in mm)

The chamber excitation was provided by a signal generator, a power amplifier and the source. The microphone was supported by a thin rod, placed in each measurement point inside the cavity. The pressure oscillations inside the cavity were captured by the microphone and registered by the data acquisition system. The volume acceleration source channel was settled as the reference channel. A white noise was used as excitation signal and the FRF were acquired at 7 points along the longitudinal axis (Fig. 24), being the point 0 the FRF driving point. The FRF were obtained by considering the volume acceleration as the excitation and the sound pressures as the responses. In order to make compatible theoretical x experimental comparisons, the volume velocity was assessed (instead of volume acceleration).

Hr Design

The objective is to tune the HR resonance as the same frequency that must be attenuated. Due to construction facility; it was chosen a cylindrical shape to develop the HR. Not only the resonance frequency must be observed during the design process, but also several factors that influence directly the behavior of the HR:

- Resonance frequency of interest = 730 Hz (second longitudinal mode to be attenuated);
- Relation HR dimensions and wavelength λ (must be at least 10 times smaller than λ);
- Area Ratio (AR) must be smaller than 0.16, in order to assure an end mass correction
- Quality Factor (Q): monitored to be used in future designs (compare with other HR shapes);
- Constructive factors. Define such dimensions that can be feasible constructively.

The parameters used in the design were updated considering the room temperature (28 °C) observed during the experiment:: sound speed: $c = 348.3$ m/s; air density: $\rho = 1.1839$ kg/m^3; air dynamic viscosity: $\mu = 1.983x10^{-5}$ kg/ms.

Considering the geometric dimensions in Eq. (14), one obtains the tuned resonance frequency as *726 Hz*, with a quality factor of *Q = 372.8*. The acoustic parameters were calculated $M = 1019$ kg/m^4, $C = 4.72x10^{-11}$ s^{2}m^4/kg, $R = 12468$ kg/sm^4, using the Eq. (9), (10) and (11), respectively. Three HR were manufactured in nylon. The presented measurement methodology was repeated in order to acquire the same FRF, considering the new configuration, with the resonators.

Results

The identified natural frequencies are summarized in Table 6. The transversal modes frequencies were identified only numerically. Considering the first four longitudinal modes, the maximum error comparing the numerical and experimental estimation of natural frequencies was 6.6%.

Figures 25a, 25c and 25e present the three first longitudinal numerical modes.

Table 6. Experimental and Numerical Natural Frequencies (Hz)

Mode	Experimental	Numerical	Error (%)
Longitudinal 1st	164.5	170.28	3.51
Longitudinal 2nd	730.85	721.9	1.23
Transversal 1st	-	1126	-
Transversal 2nd	-	1126	-
Longitudinal 3rd	1272.64	1296	1.84
Transversal 3rd	-	1382	-
Transversal 4th	-	1382	-
Longitudinal 4th	1689.52	1801	6.6

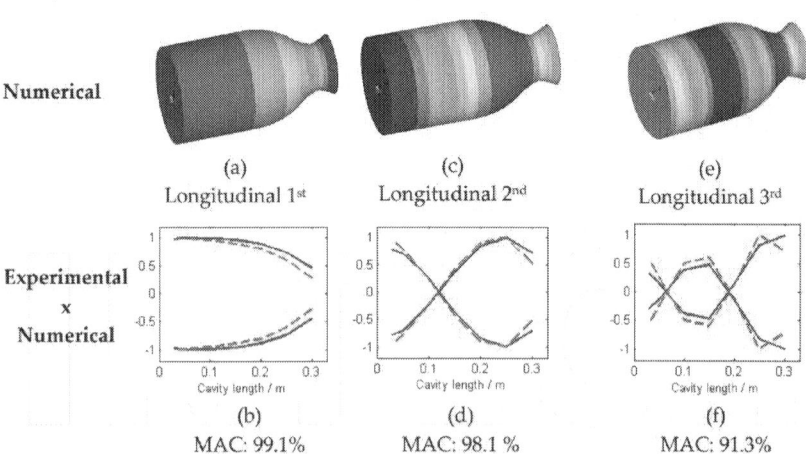

Figure 25. a) to f) Numerical modes and comparison between experimental (blue line) and numerical (red line) longitudinal mode shapes

Experimental versus numerical modes comparison, considering the normalized amplitude is also shown (Figs. 25b, 25d and 25f). In the numerical modes represented by the collor maps pictures (Figures 25a, 25c and 25e) the nodal regions are in green. The MAC (Eq. (7)) is also presented.

Notice in Fig. 25 that MAC values are bigger than 91% for the three first modes. For the first and second modes MAC values reach about 99%. Figures 25d and 25f show that the nodes in these modes are almost at the same point. After the introduction of the HR, the attenuation of the second mode is clearly noted in Fig. 26a, when comparing measurement results of the original cavity. At least 9 dB of attenuation can be observed in the new configuration. The FRF with and without HR are almost the same, but the second mode region (about 730 Hz), where the HR is tuned.

Figure 26 depicts the behavior of the chamber with HR. Numerical mode shapes of the configuration with HR were plotted and correlated to each part of the experimental FRF. The Fig. 26b was zoomed from the squared highlight in Fig.26a (point 1). This allows visualize the entire system behavior.

In Fig. 26 it can be noticed that mode shapes *(c)* and *(d)* have similar behavior, but different natural frequencies: it can be realized that the pressures inside all resonators vary in phase. In this case, the pressure inside the chamber remains almost unchanged. In the mode shapes *(e)* and *(f)*, the resonators act close its tuned frequency. In these modes, the whole chamber behaves as a nodal region and the pressure inside the resonators varies out of phase.

As a result of the movement of the air mass inside the neck in the four modes represented in Figs. 26, the acoustic energy on the resonators behaves as expected, reducing the energy inside the chamber.

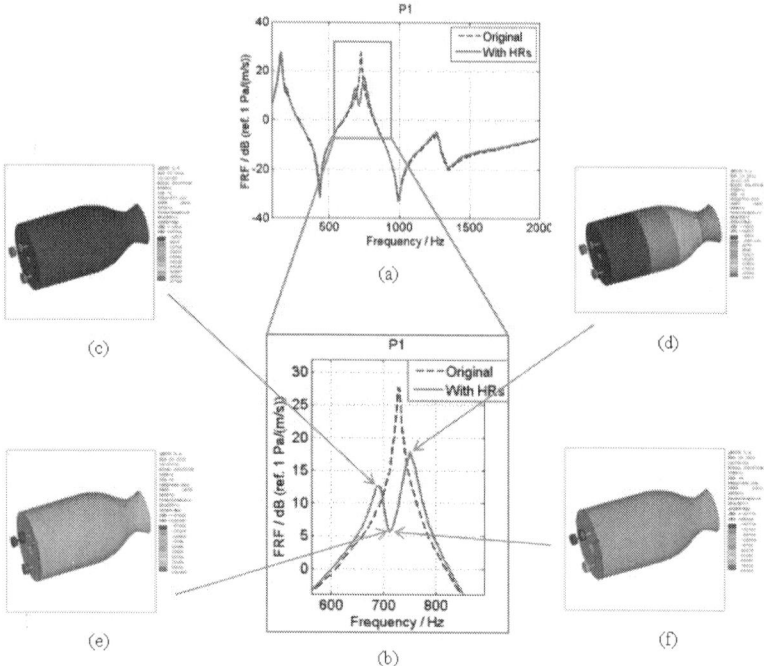

Figure 26. Numerical mode shapes near the resonances of the HR

CONCLUSIONS

This chapter presents three acoustic case studies applied on rocketry design.

Firstly, microphone protection devices design procedures for in-flight measurements are described. The modeling techniques using analytical and FEM numerical tools are presented as well as the validation acoustic testing procedures are presented. Good agreement among numerical and experimental results was obtained. A procedure to asses the SPL outer the launcher's structure by using the adapters' acoustic transfer functions and internal SPL measurements was also described.

Vibro-acoustic virtual prototypes were used to predict the acoustic response of a PLF cavity when excited by an ADF of 145 dB OSPL ranging from 5 to 8,000 Hz, generated at lift off.

Coupled deterministic techniques, using FEM/FEM and FEM/BEM, were applied to the fairing problem in a low-frequency band, considering accurate and efficient modeling techniques. The modal and semi-modal superposition techniques were applied to perform a FRA. In the higher frequencies, SEA coupling technique was applied to obtain the fairing acoustic responses in 1/3 octave bands.

The fairing was submitted to the lift off excitation in acoustic reverberant test and the internal acoustic pressure levels were measured. Experimental and numerical results show good agreement, except for the frequencies below 50 Hz and above 4,000 Hz.

The sensitivity analysis of acoustic blankets showed to be an effective tool for the development of the fairing NCT design. The effectiveness of a NCT considering its weight and performance can easily be evaluated using SEA, still in the development phase, when detailed subsystems are not required.

By analyzing many NCT configurations one can provide a library of performances and weights, important parameters that describe the ELV performance book. As in space industry the cost of a mission is a major issue, a trade-off between NCT weight and efficiency must be accounted. Acoustic testing in reverberant chamber may be conducted to validate the presented results and other porous-elastic materials may be investigated to complement the fairing NCT design library.

Finally, for the third case study, the use of a volumetric source in Experimental Acoustic Modal Analysis has important role in the process, once allows the accurate measurement of acoustic FRF.

The numerical model results were used as the basis for the HR design, in a first moment. In addition, numerical and experimental models were used to identify and localize, with a level of security, the node and maximum amplitude regions of each mode. The HR design seemed to be adequate, once it was verified an attenuation of 9 dB or bigger, depending of the location inside the chamber

REFERENCES

1. P. Jorge, Ravi. N. Arenas, Margasahayam, 2006 Noise and Vibration of Spacecraft Structures, Ingeniare. Revista Chilena de Ingeniería, 14 Nº 3, 251264

2. R. Pirk, P. Sas, W. Desmet, L. C. S. Góes, ". Vibro-Acoustic, of. Analysis, Vehicle. the, Launcher´s. . V. L. S. Sattelite, Ph. Fairing", D. Thesis, 2003 Technological Institute of Aeronautics (ITA), São José dos Campos, Brazil.

3. R. Pirk, L. C. S. Góes, 2006 "Acoustic Theoretical x Experimental Comparison of the Brazilian Satellite Launcher Vehicle (VLS) Fairing", Proceedings of the ISMA 2006, Leuven, Belgium.

4. R. Pirk, Carlos. Souto, A. d´, 2009 Implementation of Acoustic Blankets to the VLS Fairing- A Sensitivity Analysis Using SEA, Proceedings of the ISMA 2008, Leuven, Belgium.

5. J. M. Scott, B. G. Jay, C. B. Robert, 1996 General Environmental Specification for STS and ELV- Payloads, Subsystems and Components, GEVS-SE, Rev. June 1996, NASA Goddard Space Flight Center, Maryland, 20771.

6. Souto and Pirk, 2009 Estimation of the external sound pressure levels generated during VLS- 1 lift off, Institute of Aeronautics and Space, Technical Report RT 001/AIE/R/2009).

7. Pirk et al. 2010 Alternative External Acoustic Loads Estimation of the Brazilian Satellite Launcher (VLS-1), Using In-Flight Experimental Data, International Congress on Sound and Vibration, Rio de Janeiro, Brazil

8. R. M. Glaese, E. H. Anderson, 2003 "Initial Structural-Acoustic Modeling and Control Results for a Full-Scale Composite Payload Fairing for Acoustic Launch Load Alleviation". http://www.csaengineering.com/techpapers/technicalpaperspdfs/CSA1999_ Initial Structural

9. K. Weissman, M. E. Mc Nelis, W. D. Pordan, 1994 "Implementation of Acoustic Blankets in Energy Analysis Methods with Application to the Atlas Payload Fairing, Journal of the Institute of Environmental Sciences- IES, July/August 1994.

10. J. S. Bolton, 2005 "Porous Materials for Sound Absorption and Transmission Control", Proceedings of the 2005 Congress and Exposition on Noise Control Enginnering, INTERNOISE 2005, Rio de Janeiro, Brazil.

11. H. Defosse, M. A. Hamdi, ". Vibro-acoustic, of. study, V. Ariane, during. launcher, Proc. lift-off", of Ariane V launcher during lift-off", Proc. of Inter-Noise 2000 1 9-14 (Nice, 2000).

12. F. E. C. Culick, Combustion Instabilities in Liquid Rocket Engines: Fundamentals and Control. California Institute of Technology, 2002

13. Burnley, V.S.; Culick, F.E.C., The Influence of Combustion Noise on Acoustic Instabilities, Air Force Research Laboratory, OMB No 0704-0188, 1997.

14. R. Pirk, C. d´, A. Souto, D. D. Silveira, C. M. Souza, 2010 Liquid rocket combustion chamber acoustic characterization, Journal of Aerospace Technology and Management.

15. E. Laudien, Others, 1994 Experimental Procedures Aiding the Design of Acoustic Cavities, DASA- Deutsche Aerospace AG, Liquid Rocket Engine Combustion Instability, Progress in Astronautics and Aeronautics, Chapter 14, 169

16. M. S. Natanzon, 1999 Edited by Culick, F. E. C., California Institute of Technology

17. Troclet, B., Analysis of Vibro-acoustic Response of Launchers in the Low-Frequency and High-Frequency Range, Proc. of NOVEM 2000 (Lyon, France, 2000).

18. V. I. Khlybov, S. A. Mak´hankov, 2009 The Preparation to Experimental Studies and Acoustic Sensors Testing Program Development, Technical Report.

19. Lawrence. E. Kinsler, Austin. R. Frey, of. Fundamentals, John. Acoustics, Wiley, 2nd. Sons, edition, 1962EO

20. W. Desmet, D. Vandepitte, 2001 Finite Element Method in Acoustics, Course Graduate School in Mechanics- Advanced Acoustics, Katholieke Universiteit Leuven

21. J. Coyette-P, C. Lecomte, K. Meerbergen, 1997 Treatment of Random Excitations using SYSNOISE Rev. 5.3.1 - Documentation Theoretical Manual, Users Manual and Validation Manual, LMS Numerical Technologies NV.

22. J. Klos, et, 2003 Sound Transmission Through a Curved Honeycomb Composite Panel, AIAA Conference Paper AIAA-20033

23. Desmet, W, Vandepite, D., Boundary Elements in Acoustics, Katholieke Universiteit Leuven (KUL), 2001.

24. Lyon and DeJong, Theory and Application of Statistical Energy Analysis, Second Edition, 1995

25. T. Thinh, Modeling. Vibro-acoustic, of. Study, Delta. I. I. 1. the, 10 10 Composite Fairing, Journal of the IEST, November/December 1999.

26. J. F. Allard, 1993 Propagation of Sound in Porous Media- Modeling Sound Absorbing Materials, Elsevier Applied Science, London and New York.

27. G. P. Sutton, O. Biblarz, 2001 "Rocket Propulsion Elements", New York, John Wiley & Sons.

28. Santana Jr, A. , Investigation of Passive Control Devices to Suppress Acoustic Instability in Combustion Chambers. Thesis of doctor in science. Aeronautics Institute of Technology, São José dos Campos, Brazil, 2008

29. G. P. Guimarães, R. Pirk, C. A. Souto, L. C. S. Góes, Acoustic Modal Analysis of Cylindrical-Type Cavities. Proceedings of 8 8th Intern. Confer. on Structural Dynamics- EURODYN, Leuven, Belgium, 2011, 31603167 .

30. F. Fahy, Fundamentals of Engineering Acoustics. Academic Press. London, UK. 2001

CITATION

Rogério Pirk, Carlos d´Andrade Souto, Gustavo Paulinelli Guimarães and Luiz Carlos Sandoval Góes (2013). Acoustics and Vibro-Acoustics Applied in Space Industry, Modeling and Measurement Methods for Acoustic Waves and for Acoustic Microdevices, Prof. Marco G. Beghi (Ed.), ISBN: 978-953-51-1189-4, InTech, DOI: 10.5772/49966.

Chapter 3

Surface Acoustic Wave Based Magnetic Sensors

Bodong Li[1], Hommood Al Rowais[2] and Jürgen Kosel[1]

[1]Electrical Engineering Department, King Abdullah University of Science and Technology, Thuwal, Saudi Arabia
[2]Electrical Engineering Department, Georgia Institute of Technology, Atlanta, Georgia, USA

INTRODUCTION

Since the radar system was invented in 1922, the development of devices communicating by means of reflected power has experienceda continuously growing interest. In 1948, Harry Stockman published a paper [1] in which he laid the basis for the idea of radio frequency identification (RFID), and the first patent had been filed in 1973 by Charles Walton. After decades of research and commercialization, RFID products became a part of everyday life (e.g. logistics, access control, security). With the growing interest in remote and battery-free devices, researchers are pushing the boundaries of RFID technology to find solutions in new fields like sensing applications.

For many sensors such as those operated in remote or harsh environments, the sensitivity is not the only evaluation criteria. The lifetime, especially of the power source, and the complexity added by wiring often demand wireless and passive operation.Batteries have limited lifetime and also add to the size and mass of the sensors. Alternatively, energy harvesting oran RF-based wireless power supply can be employed [2, 3]. The former method depends on environmental conditions such as solar radiation, temperature change, chemical reagents, vibration etc., which are often not constantly or not sufficiently available.RF power sources, on the other hand, transmit power wirelesslyand with full control over amount and timing.

Passive and remote sensors utilizing SAW transpondersaredevices, which are powered by an RF source. These systems require an interrogation device that requests the sensor signal, a SAW transponder plus a sensing element and two antennas. The basic idea is that an RF signal of certain frequencies generated by the interrogator is received by the SAW transponder, which reflects back a signal modified by the sensing element. This signalcontains the environmental information in an amplitude and phase change, which is converted into the physical parameters by the interrogator. In most cases, SAW sensors are coded by having different reflector designs in order to have multiple measurement capabilities from sensors located in the same interrogation area. A great amount of research has been carried out in the past decades in this field, and, as a result, different wireless SAW sensors have been developed to measure a variety of physical and chemical parameters including temperature, stress, torque, pressure, humidity, magnetic field, chemical vapor etc. [4-9]. Several devices are already commercialized [10-12].

SAW-based magnetic sensors have, so far, not been studied in detail. Magnetic sensors are one of the most pervasive kinds of sensors for a large number of applications and are employed in different fields like automotive, biomedical or consumer electronics. Integrating magnetic sensors with SAW transponders enables remote and passive operation, thereby, opens a door for further applications. Early in 1975, a magnetically tuned SAW phase shifter was proposed by Ganguly et al [13]. A thin film of magnetostrictive material was fabricated on the delay line of a SAW device. A phase shift was observed due to the dependence of the wave propagation velocity on the external magnetic field.Recently, a new concept of amagnetic sensor based on a SAW resonator has been published [8]. A magnetostrictive material was used to fabricate the interdigital transducers (IDT) of the SAW device. The resonant frequency of the device changes with an external magnetic field. A different idea was put forth by Hauser and Steindl[14-16] combining a SAW transducer with a giant magnetoimpedance (GMI) microwire sensor. The GMI sensor has a magnetic sensitivity, at least, one order higher than contemporary giant magnetoresistance (GMR) sensors and can be used to measure very low magnetic fields such as those generated by the human heart or muscles. A GMI sensor is operated by an accurrent and the impedance changes upon changes of a magnetic field. This makes it a suitable load for a SAW transponder, which converts this impedance change into a magnitude and phase change of the reflected acoustic waves. In order to reduce the size and improve the level of integration of the senor, a new design of an integrated SAW transponder and thin film GMI sensor has been proposed and developed recently by the authors [17, 18]. The SAW transponder and GMI thin film were integrated on the same chip using standard micro-fabrication technology suitable for mass fabrication.

The ideal SAW-based magnetic sensor is small and highly integrated, inexpensive, passive, remotely controlled and have a high magnetic sensitivity together with a large linear range. With regard to these criterions, an SAW-GMI sensor is avery promising candidate.

SAW-based magnetic sensors have been studied for several years.However, this topic has yet not been comprehensively summarized, and the aim of this chapter is to provide a systematic review of the past research as well as the latest results. The performance of the devices crucially depends on different design parameters in a complex fashion. This will be shown by a detailed description and analysis for a device consisting of a SAW transponder and GMI thin film sensor.

SAW BASED PASSIVE SENSORS

A basic SAW device consists of an input interdigital transducer (IDT) and an output (or reflector) IDT, which are fabricated on a piezoelectric substrate. The area between the input IDT and output IDT is called the delay line. The IDT is made of two metallic, comb-like structuresarranged in an interdigital fashion, whereby the distance between two fingers of a comb defines the periodicity (p) (Fig. 1). Upon application of a voltage, charges accumulate at the fingers of the IDT depending on the capacitance of the structure.The resulting electric field generates stress in the substrate due to the piezoelectric effect.If an ac input voltage is applied, the continuously changing polarity of the charges will excite an

SAW (Rayleigh wave) traveling through the substrate. At the operating (resonant) frequency of the SAW device,the value of$p$ equals the wavelength of the SAW, and the SAW amplitude showsa maximum value due to constructive superposition.

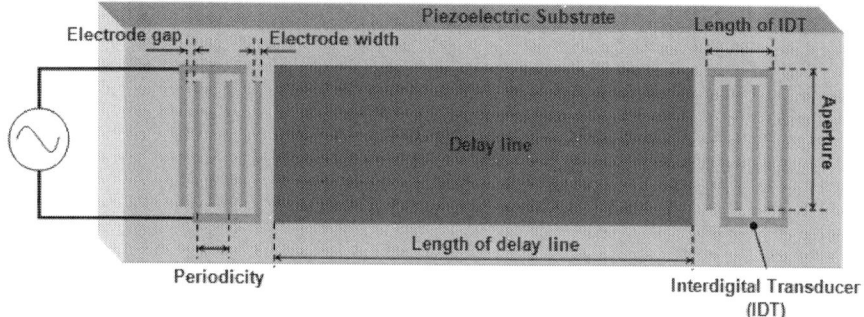

Figure 1: Schematic of a SAW device.

Basics of Saw Devices

SAW Resonant Frequency:The resonant motion of an acoustic resonator is caused by the coupling between the transducer (IDT) and the acoustic medium. The resonant frequency, or operating frequency, is determined by the periodicity and the acoustic wave velocity(v)

$$f = \frac{v}{p}.$$

(1)

The value of $v$mainly depends on the substrate's material. A typical SAW velocity for piezoelectric materials is several thousand meters per second. Due to the intrinsic anisotropy of piezoelectric materials, v is dependent on the direction of propagation. Since different acoustic modes have different wave velocities, a device can resonate at different frequencies. The SAWs are Rayleigh waves, which have a longitudinal and a vertical shear component that can couple with any media in contact with the surface. This coupling strongly affects the amplitude and velocity of the wave allowing SAW sensors to directly sense, e.g., mass loads.

Electro-Mechanical Coupling Coefficient: The electro-mechanical coupling coefficient (κ) defines the conversion efficiency of the piezoelectric material between the electrical and mechanical energies, determined by

$$\kappa^{-2} = \frac{input\ energy}{converted\ energy}.$$

(2)

A high coupling coefficient reduces the insertion loss caused by the energy conversion, which results in smaller energy consumption as well as larger effective readout distance of a SAW-based wireless sensor.

SAW Delay Line: The SAW delay line refers to the area between the input IDT and output IDT on the substrate (Fig. 1). It creates a time delay between the input signal and the output signal depending on the SAW velocity and the length of the delay line. Due to this feature, SAW devices are widely used in RF electronics. It is also used in sensing applications, where the measurand causes, e.g., a change in the SAW.

Temperature Coefficient of Delay (TCD): The TCD reflects the temperature dependence of the time delay and is connectedwith the thermal expansion coefficient (α) and the temperature coefficient of the phase velocity (TCV) by

$$TCD = \alpha - TCV .$$

(3)

The temperature dependence of the time delay is the basis of SAW temperature sensors, where higher TCD values yield higher sensitivity. However, for other SAW devices, the influence of the TCD on the time delay is undesirableand has to be minimized or eliminated. For this purpose, temperature compensated cuts of the crystalline substrates are employed, where the TCD is minimized over certain temperature ranges [19-20]. Piezoelectric bi-layers are another concept that has been utilized in order to compensate the TCD in sensing applications [21, 22].

Basic Design Concepts of Passive Saw Sensors

Passive SAW sensors typically operate asresonators, delay lines or loaded transponders. In case of resonators, the reflection of the interrogation signal from the SAW device is a function of the SAW device's resonant frequency, which depends on the measurand. In case of delay lines, the request signal is separated from the response signal by a time difference, whereby this time difference depends on the measurand. Similarly, the request signal and response signal are separated by a time difference in case of a loaded transponder. However, the time difference is constant and the measurand affects the signal amplitude. Intrinsic SAW sensors utilize a change of the substrate's properties. For example, intrinsic temperature sensors were realized by detecting the resonant frequency or phase change of the SAW in materials with large TCD [23]. Intrinsic stress sensors utilize the length change of the delay line caused by mechanical strain applied to the substrate. The stress can be evaluated by measuring the SAW phase shift [24]. Extrinsic SAW sensors can be realized by integrating a SAW device and an additional sensing element. A common extrinsic sensor conceptutilizes selective thin films on top of the delay line leading to a change in mass by the measurand [13, 25]. This can be, for example,a thin film with high CO_2 solubility and selectivity [26]. As CO_2 dissolves into the film, the additionalmass loadcausesa detectable phase shift in the SAW. Another extrinsic concept utilizes a sensitive IDT. For example, in case of a magnetostrictive IDT, a magnetic field applied to the sensor causes a change of the resonant frequency [8]. A loaded transponder is another extrinsic design, where the output IDT is connected to a sensor, which changes the IDT's electrical characteristics as a function of the measurand.An example for a load sensor is a pair of conducting rods placed in the earth with a certain distance from each other. As the water level changes, the resistance between the rods changes, which can be detected as a magnitude and phase change of the signal reflected from the load IDT [27]. Another example for a load is a giant magnetoimpedance sensor [16, 17]. Achange in the magnetic field yields a change in the sensor's impedance. Consequently, changes the reflectivity of the output IDT.

Some SAW sensors, their classification and method of detection are presented in Table 1.

Table 1: SAW-based passive sensors.

Sensor Type	Commercialization	Year	Intrinsic/Extrinsic	Design	Detection Method	Access	Paper
Temperature	Yes	1990	Intrinsic	Resonator	Frequency	None	[23]
		2003	Intrinsic	Delay line	Phase velocity	None	[28]
Pressure	Yes	2001	Extrinsic	Loaded Transponder	Phase	Capacitive pressure sensor	[25]
		2007	Intrinsic	Delay line	Phase	None	[24]
Bio/Chem	No/Yes	2006	Extrinsic	Resonator	Frequency	Thin film	[29]
		2011	Extrinsic	Delay line	Phase	Thin film	[26]
		2001	Extrinsic	Loaded Transponder	Amplitude/Phase	Conducting rods	[30]
Magnetic	No	1975	Extrinsic	Delay line	Phase	Thin film	[13]
		2011	Extrinsic	Resonator	Frequency	Magnetostrictive IDTs	[8]
		2006/11	Extrinsic	Loaded Transponder	Amplitude	GMI wire/ thin film	[16,17]
Sound	No	2005	Extrinsic	Loaded Transponder	Phase	Capacitive pressure sensor	[31]
Torque	Yes	1996	Intrinsic	Delay line	Phase	None	[32]

PASSIVE AND REMOTE SAW-BASED MAGNETIC SENSORS

Magnetic sensors are one of the most versatile sensors employed not only for the task of measuring magnetic fields but for a large number of different applications, thereby detecting the measurand indirectly, e.g., via a change of material parameters in construction monitoring or a change of distance in position monitoring. A passive and remote operation of magnetic sensors can be advantageous in many cases and considerably increase their applicability.

A SAW-based passive magnetic sensor can be realized either by adding an additional material layer, which is sensitive to magnetic fields, or by loading the output IDT with a magnetic sensor. In the first case, the magnetic layer changes the delay line or the resonant frequency of the SAW device. While in the second case, the sensor changes the reflection signal of the output IDT. Since SAW devices are operated by RF power, the sensor element has to work at the operation frequency of the SAW device. Among the available magnetic sensors, GMI sensors are the most suitable candidates as they have a high magnetic sensitivity as well as a high operating frequency.

Magnetostrictivesaw Devices

Magnetostriction defines the relationship between the strain and the magnetization states of a material. It is an important property of ferromagnetic materials and was first observed by James Joule in 1842 in nickel samples. For a positive/negative magnetostrictive material, an applied magnetic field causes the material to expand/shrink in the field direction. Inversely, when a stress is applied to the magnetostrictive material, its magnetic anisotropy will change accordingly.

A magnetostrictive-piezoelectric resonator consists of amorphous magnetostrictive material layers as the electrodes sandwiching a piezoelectric core (Fig. 2). An ac signal applied to the electrodes causes the quartz layer to oscillate. The resonant frequency of this oscillation depends on the thickness of the piezoelectric material, the crystal orientation, temperatureand mechanical stress, etc.

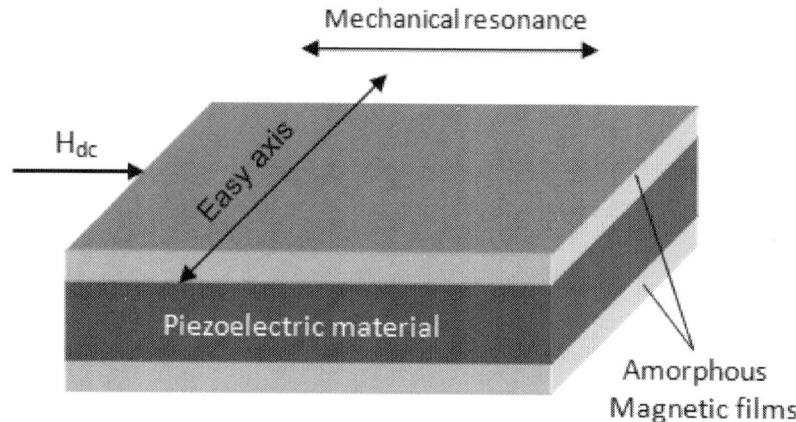

Figure 2. Structure of a compositemagnetostrictive/piezoelectric resonator.the magnetic anisotropy is perpendicular to the external magnetic field *Hdc*.

When a magnetic field*Hdc* is applied, the length change induced in the magnetostrictive film exerts stress to the piezoelectric material and, consequently, shifts the resonant frequency of the device. Utilizing this concept,a magnetic sensitivity high enough to detect the terrestrialfield has been achieved [33].In a similar work, a magnetostrictive-piezoelectric tri-layer structure has been embedded in a coil. The dc magnetic field sensitivity was as high as 10^{-8} T [34].

Passive Resonator

A SAW-based, passive resonator for magnetic field detection was developed recently by Kadota et al [8]. Nickel, which is a negative magnetostrictive material, was used to fabricate the sensing IDT on a quartz substrate (Fig. 3). Upon the application of a magnetic field, stress will be induced to the substrate by the IDT change causing a change in the resonant frequency. This sensor showed a frequency change of 200 ppm for a magnetic field of 100 mTapplied perpendicularly to the direction of SAW propagation.

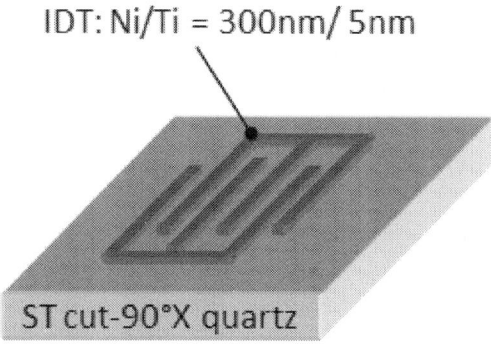

Figure 3. Schematic of amagnetic sensor device using a magnetostrictive IDT on a SAW substrate.

Passive Phase Shifter

A magnetically tuned SAW phase shifter is a one-port SAW structure with a magnetic sensing functionality achieved through a delay line sensitive to magnetic fields. This idea was first introduced by Ganguly et al in 1975 [13]. In their device, the acoustic velocity is varied by an external magnetic field. This functionality is facilitated by a magnetostrictive thin film deposited on top of the delay line (Fig. 4). The propagation velocity of the SAW in the film region depends on the magnetic field. Hence, there is a correlation between the time shift of the reflected signal and the magnetic field.

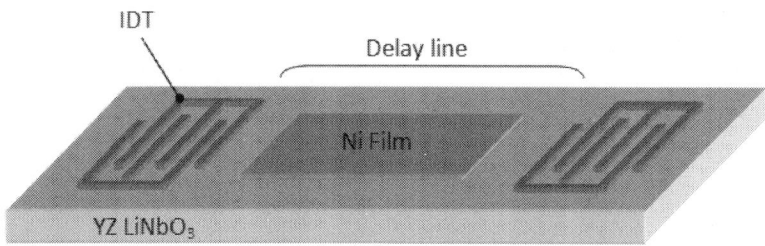

Figure 4. Schematic of a magnetically tuned SAW wave phase shifter.

Later, research efforts focused on different magnetostrictive materials and measurement methods [9,35, 36], and a magnetic sensitivity of 10^{-4}%/kOe was achieved.

Saw-Gmi Sensors

A SAW-based, magnetic and passive sensorscomprises a two-port SAW transponder and a magnetic sensor acting as a load at the output IDT. Among the available magnetic field sensors, giant magnetoimpedance (GMI) sensors offer favorable characteristics like high sensitivity to magnetic fields and high operation frequency (compatible with SAW transponders) making them a very suitable load. SAW-GMI sensors have been fabricated by combining SAW transponders with GMI wire sensors as well as thin film GMI sensors. Both of these methods have shown a higher magnetic sensitivity than direct designs.

Basics of GMI Sensors

The GMI effect was first observed in Co-based amorphous wires by Panina and Mohri in 1994[37] and has since attracted strong interest due to its sensitivity enabling magnetic field measurement with a nT resolution. The GMI effect isthe impedance change of an ac-powered ferromagnetic conductor upon the change of a magnetic field. The relative impedance change, also called GMI ratio, is expressedas

$$\text{GMI Ratio (\%)} = 100\% \times \frac{Z(H)-Z(H_0)}{Z(H_0)} \quad \text{or} \quad \text{GMI Ratio (\%)} = 100\% \times \frac{Z(H)-Z(H_{max})}{Z(H_{max})}, \tag{4}$$

where$Z(H0)$ is the impedance at zero magnetic field and$Z(Hmax)$ is the impedance atsaturation field. Both definitions have particular aspects that should be considered. In case of the first expression,$Z(H0)$depends on the remanent state of the magnetic material while,in the second case, $Z(Hmax)$ is not always achievable and equipment dependent.

The GMI effect is explained by classical electromagnetism.The change of the complex impedance mainly originates from the skin effect in conjunction with a change of the complex permeability.Analytically, the complex impedance (Z) of a conductor is defined by

$$Z = \frac{U_{ac}}{I_{ac}} = \frac{\int_L \frac{1}{\sigma} J_z(S)dz}{\iint_q J_z dq}, \tag{5}$$

where$Uac$is the applied ac voltage, $Iac$is the current, L is the length and σ the conductivity.$S$and $q$refer to the surface and the cross section of the conductor, respectively.$Jz$is the current density in the longitudinal direction obtained by solving Maxwell's equations.In ferromagnetic materials, by neglecting displacement currents ($D'=0$), Maxwell's equations can be written as follows:,

$$\nabla \times H = J, \tag{6}$$

$$\nabla \times J = -\frac{\mu_0}{\rho_f}(\dot{H} + \dot{M}), \tag{7}$$

$$\nabla \cdot (H + M) = 0, \tag{9}$$

J is the current density,H is the applied magnetic field, M is the magnetization of the ferromagnetic material, $\rho_f$is the free charge density and $\mu_0$is the permeability of vacuum.From Equ.(6) to (8), the expression

$$\nabla^2 H - \frac{\mu_0}{\rho_f}\dot{H} = \frac{\mu_0}{\rho_f}\dot{M} - \nabla(\nabla \cdot M), \tag{9}$$

can be derived. Equ. (9) can be solved using the Landau-Lifshitz equation, whichrelates $M$and$H$

$$\dot{M} = \gamma M \times H_{eff} - \frac{\alpha}{M_s}M \times \dot{M}, \tag{10}$$

Where γ is the gyromagnetic ratio,$M_s$is the saturation magnetization,α is the damping parameterand$H_{eff}$is the effective magnetic field expressed as [30]

$$H_{eff} = H + H_a + \frac{2A}{\mu_0 M_s}\nabla^2 M \tag{11}$$

Where H is the internal magnetic field that includes the applied field and demagnetizing field, $H_a$is the anisotropy field and$A$ is the exchange stiffness constant..

By combining Equ.(5) to (11), a theoretical impedance model can be evaluated for GMI sensors with different geometries [37-40].

Although the experimentally obtained GMI effect shows a large sensitivity compared to other effects exploited for magnetic sensors, the theoretically estimatedvalues have not been achieved yet. Therefore, a lot of effort has been putinto improving the magnetic properties of GMI materials [41-44]. At the same time, GMI sensors of different structures have been developed such as glass-coated wires, thin films, multi layer thin films, meander structures, ribbons, etc. [45-47]

Wire GMI Sensors

As the first discovered GMI sensor structures, GMI wire sensors have been extensively studied. Based on the classical electromagnetism, the theoretical model of the GMI wire is (Panina et al, 1994) [37]

$$Z = \frac{R_{dc}kr\zeta_0(kr)}{2\zeta_1(kr)},$$

(12)

where

$$k = \frac{1+j}{\delta m}$$

(13)

and

$$\delta m = \frac{c}{\sqrt{4\pi^2 f \sigma \mu_{\emptyset}}}.$$

(14)

R_{dc} is the dc resistance of the wire, $\zeta 0, \zeta 1$ are the Bessel functions, r is the radius of the wire, j is the imaginary unit, δm is the penetration depth, c is the speed of light, f is the frequency of the ac current, $\mu_{\emptyset}$ is the circumferential magnetic anisotropy. The origin of the GMI effect lies in the dependence of $\mu_{\emptyset}$ on an axial magnetic field resulting in a change of δm. In order to obtain a high GMI ratio, the value of δm has to be close to the thickness of the conductor. Hence, the thinner a ferromagnetic conductor and the lower its permeability, the higher the operation frequency required. A well-defined circumferential magnetic anisotropy in combination with a soft magnetic behavior is desirable, since it will provide a large permeability change for small magnetic fields.

Different amorphous and ferromagnetic materials were used to fabricate GMI wires [48], and various fabrication methods were developed such as melt spinning, in-rotating water spinning, glass-coated melt spinning etc. [45, 49, 50].Glass-coated micro-wires (Fig. 5)present outstanding properties in terms of the magnetic anisotropy distribution, which is reinforced by the strong mechanical stress induced by the coating. $(Co_xFe_{1-x})_{72.5}Si_{12.5}B_{15}$is one of the most typical materials. By adjusting x from 0 to 1, the magnetostriction of the material changes from positive at high Fe content to negative at high Co content. Negative magnetostrictive compositions in combination with the compressive, radial stress induced by quenching and the glass coating provide the best results, since it supports a strong circumferential anisotropy.

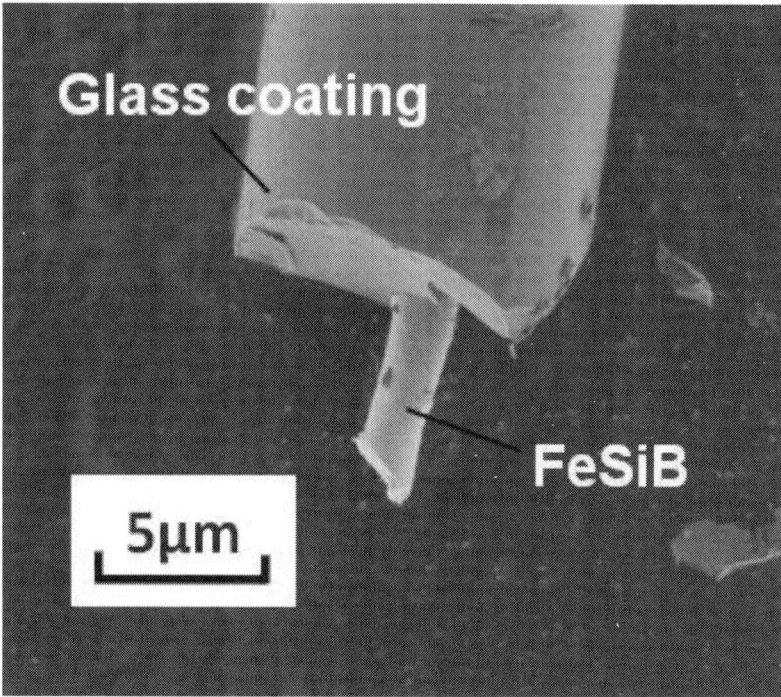

Figure 5. SEM image of a glass-coated amorphous micro-wire (Courtesy of M. Vazquez, Inst. Materials Science of Madrid, CSIC).

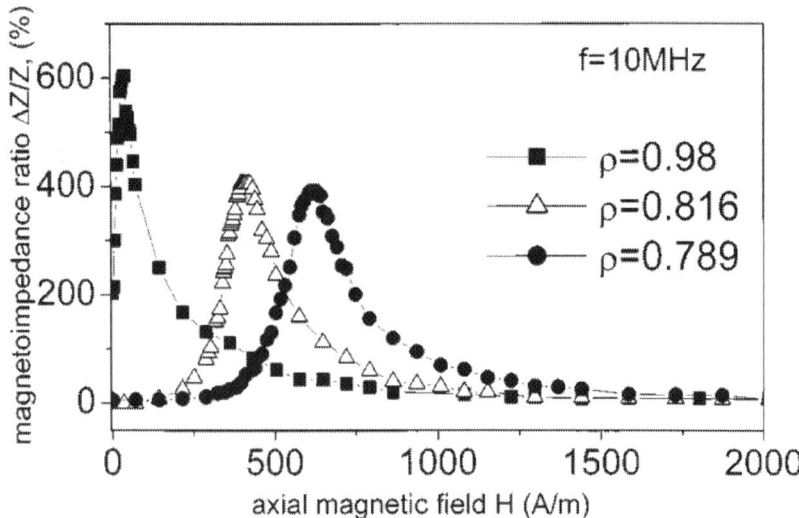

Figure 6. GMI ratio of Co67Fe3.85Ni1.45B11.5Si14.5Mo1.7 glass-coated wires with different geometric ratioρ(themetallic nucleus diameter to the total microwire diameter)at 10 MHz.

Wire-type GMI sensorsprovide the best performance in terms of the GMI ratio with values as high as 615% (Fig. 6)achieved with optimized glass coated microwires (Zhukova et al, 2002) [43].The value of the magnetic field at which the maximum GMI ratio is obtained increasesas the diameter of the magnetic nucleus decreases compared to the diameter of the glass coating. This is attributed to the different anisotropies induced by the stress from the coating. Due to the high sensitivity provided by GMI wiresthey have been commercialized despite the facts that fabrication is not silicon based, does not use standard microfabrication methods and, as a consequence, integration with electronics is complex.

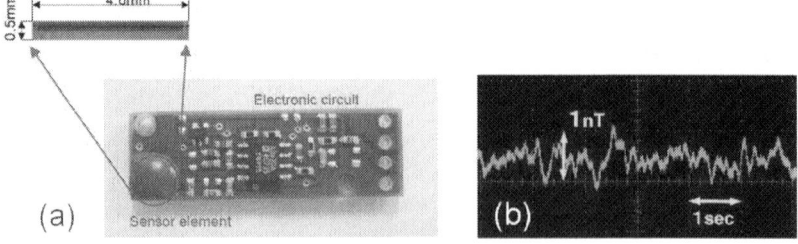

Figure 7. a) Layout of the commercialized GMI sensor from Aichi Steel Co. (b) Noise output of the GMI sensor.

Fig. 7 showsa GMI sensor developed by Aichi Steel Co., which has a very high sensitivity of 1V/μT and a noise level of 1 nT[51].

Ribbon GMI Sensors
Magnetic ribbons discussed in this section are planar structuresof rectangular shape with a thickness of a few tens of micrometers and a length and width from several millimeters to centimeters.similar to the micro-wires, magnetostriction is utilized in order to create certain anisotropies during the ribbon's fabrication. Magnetic ribbons that exhibit a strong GMI effect have a high permeability as well as a transversal magnetic anisotropy.

For a planar film of infinite width, the impedance is given by

$$Z = R_{dc} \cdot \frac{jka}{2} coth(\frac{jka}{2}) \, ,$$

(15)

whereRdc is the dc resistance, a is the thickness of the ribbon, $k$and δm can be obtained fromEqu. (14) with the only difference that $\mu\o$represents the transversalpermeability instead of the circumferential one [52].

Again, Fe-and Co-based amorphous alloys are preferably used as the magnetic material. The standard fabrication method for the ribbons is melt spinning, where a rotating copper wheel is used to rapidly solidifythe liquid alloy.This method produces magnetic ribbons with a thickness of about 25μm and a width of several mm. With this thickness, ribbon GMI sensorsoperate at comparably low frequencies of hundred kHz up to a few MHz.A GMI ratio of, e.g., 640% has been obtained with a GMI ribbon made of $Fe_{71}Al_2Si_{14}B_{8.5}Cu_1Nb_{3.5}$at 5 MHz [53].

Thin Film GMI Sensors
In theory, a single layer magnetic thin film is similar to a magnetic ribbon,and the same analytical expressions are applied for modeling the GMI effect. Practically, the main difference is the fabrication method.

Thin film fabrication is a standard micro-fabricationtechnology producing a film thickness of somenanometersup to a few micrometers. Thin film GMI sensorsare of great interest due to the advantages arising from the fabrication in terms of the flexibility in design and integration. They can easily be fabricated on the same substrate as the electronic circuit and other devices. In the context of passive and remote sensors, this is particularly relevant, since the GMI element can be easily integrated with an SAW device. For this reason, GMI thin film sensors will be discussed in more detail and our recent results will be presented.

Compared to wires and ribbons, the results obtained with thin film sensors have not been as good, and the highest GMI ratios reported are around250%[42]. This may be due to the differencesin the magnetic softness as well as the magnetic anisotropy, which is very well established in circumferential and transversal direction in wires and ribbons, respectively, and is difficult to control in thin films. Inthin films transverse anisotropy is mainly realized through magnetic field deposition or field annealing,Fig. 8 (a) and (b) show the magnetization curve and domain structure of a$Ni_{80}Fe_{20}$thin film (100nm thick) fabricated under a magnetic field of 200 Oe during deposition. A magnetic easy axis and domain structures in transverse direction are observed. Due to the small thickness, thin film GMI sensors normally operate at a higher frequency from hundred MHz to several GHz where the penetration depth is in the range of the film thickness.

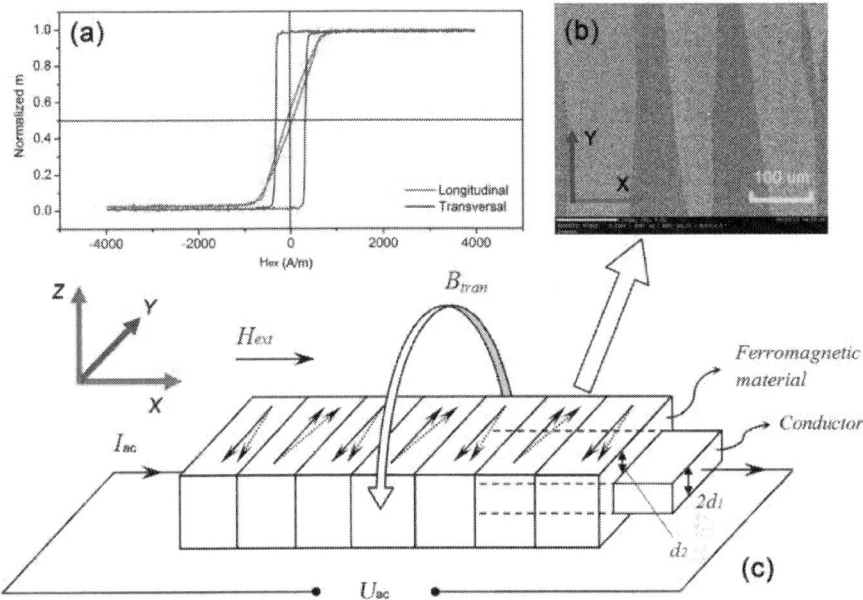

Figure 8. a) Magnetization curves obtained by vibrating sample magnetometry of a magneticthin film (100nm of $Ni_{80}Fe_{20}$) in transversal and longitudinal directions. (b) Domain pattern of the magnetic layer. (c) Schematic of a typical multi layer GMI structure. The arrows in the ferromagnetic material indicate the magnetization of individual domains (simplified). Upon application of an external magnetic field*Hext*, the magnetization rotates into the direction of *Hext* (dotted arrows).

In general, a GMI sensor with high sensitivity consists of a stack of several material layers. In case of a tri-layer element, one conducting layer is sandwiched between two magnetic layers as shown in Fig. 8 (c). The conducting layer ensures a high conductivity and, in combination with the highly permeable magnetic layers, a large skin effect is obtained [51, 54].An alternating current *Iac* mainly flowing through the conductor generates a transversal flux *Btran*, which magnetizes the magnetic layers. Upon the application of an external field *Hext*in longitudinal direction, the magnetization caused by *Iac*will be changed. This is equivalent to a change of the transversal permeability of the magnetic layers and is reflected by an impedance change.

The analytical model of the impedance for a magnetic/conducting/magnetic tri-layer structureis given by

$$Z = R_{dc}(1 - 2j\mu\frac{d_1d_2}{\delta_c^2}),$$

(16)

whereRdc is the dc resistance of the inner conductor, $2d1$is the thickness of the conductor,$d2$is the thickness of the magnetic layers as shown in Fig. 8 (c) andδc is the penetration depth of the conducting layer[39].

Analytical solutions for the impedance of thin film GMI sensors can only be found for rather simple structures. In order to calculate the impedance of more complicated geometries, for example,a sandwich structure with isolation layersbetween the conductor and the magnetic layers [41], a meander structure multilayer [46] or to take into account edge effects, the finite element method (FEM) provides a viable solution [55].

Fig. 9 shows the comparison of the GMI ratios simulated for a single magnetic layer, a tri-layerstructure made of a magnetic/conducting/magnetic stack and a five-layer structure with isolation layers between the conducting and magnetic layers using the FEM. The simulated GMI sensors have a width of w = 50 µm and length of l= 200µm. The magnetic layers have a thickness of $tmag$ = 1 µm andthe conducting layer has a thickness of$tmet$ = 4 µm. The material of the isolation layer is SiO_2 with a thickness of 1µm. The conductivity of the ferromagnetic and conducting layers are 7.69×10^5S/m (($CoFe)_{80}B_{20}$) and 4.56×10^7S/m (Gold), respectively. All parameters including Ms = 5.6×10^5A/m, γ= 2.2×10^5m/(A s), α = 0.3, Ha= 1890A/m and are taken from literature [56].

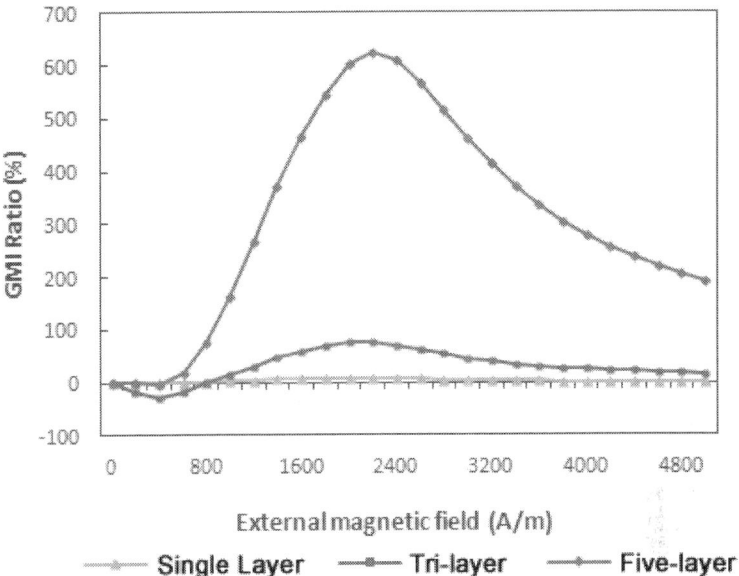

Figure 9. Simulated GMI ratios of single layer, sandwiched multilayer and isolated sandwiched multilayer structures.

The resultsclearly show the performance increase achieved with the multilayer structures. Specifically, the isolated sandwich structure has a superior performance, which is due to preventing the current from flowing in the magnetic layer.

For the fabrication of thin film GMI sensors, Co-based and Fe-based amorphous magnetic alloys were used in earlierstudies. Recently, permalloy, which is a NiFe compound, became popular as it provides very high permeability, zero magnetostriction and simple fabrication. Meander shaped multilayers and different stacks of magnetic and conductive layers using permalloy were developed. Some results are summarized in Table 2.

Table 2 Recent results on thin film GMI sensors.

Year	Material	Frequency	GMI Ratio (%)	Sensitivity (%/Oe)	Reference
1999	FeNiCrSiB/Cu/FeNiCrSiB	13MHz	77	2.8	[57]
2000	FeSiBCuNb/Cu/FeSiBCuNb	13MHz	80	2.8	[58]
2004	$Ni_{81}Fe_{19}$/Au/$Ni_{81}Fe_{19}$	300MHz	150	30	[59]
2004	$(Ni_{81}Fe_{19}$/Ag)n	1.8GHz	250	9.3	[42]
2005	$FeCuNbSiB/SiO_2/Cu/SiO_2/F$ eCuNbSiB	5.45MHz	33	1.5	[60]
2011	NiFe/Ag/NiFe	1.8GHz	55	1.2	[61]
2011	NiFe/Cu/NiFe	20MHz	166	8.3	[62]

GMI thin film sensors not only offer the advantages of standard microfabrication and straight-forward integration with SAW devices, but, as can be seen from Table 2, the operation frequency of GMI thin film sensors is also compatible with the one of SAW devices (usually from hundred MHz to several GHz and can be adjusted within a wide range.

Integrated Saw-GMI Sensor

In the first studies, SAW transponders and GMI wire sensors werecombined to form remote devices [14-16]. GMI wires were selected for their high sensitivity, and they were bonded totheoutput IDT of the SAW device, which operated as a reflector, in order to act as load impedance. The strong dependence of the impedance on magnetic fields causes a considerable amplitude dependence of the reflected signal on magnetic fields. Even though these studies provided good results for passive and remote magnetic field sensors, the fabrication method for the GMI wires, which is not compatible with standard microfabrication, is a considerable problem with respect to reproducibility and costs, hence, hindering commercial success of such devices. In order to conquer this problem, a fully integrated SAW-GMI design utilizing standard microfabrication processes is required. The most viable option is a thin film GMI sensorsfor the following reasons:

1. Thin film GMI sensors can be produced by the same metallization processes as the SAW transponders and on the same substrate.

2. Standard photolithography technique guarantees an accurate and reproducible alignment of the two devices.

3. Thin film GMI sensors provide a wide range of working frequencies up to GHz, which matches the high frequency requirement of the SAW transponders.

4. Thin film GMI sensor can have a minimized and flexible design as well as large magnetic field sensitivity.

In this section, a detailed description of our recent work on the design, fabrication and testing of an integrated SAW-GMI sensor is presented. Design

Fig. 10 shows a schematic of a GMI thin film sensor integrated with a SAW transponder. A wireless signal applied to the source IDT (IDT1) is converted to anSAWand propagates towards the other end of the substrate, where it is reflected from the reference IDT (IDT2) and the load IDT (IDT3). The reflected waves containing the reference and load information are received by IDT1 at different time instants and reconverted to a wireless electrical signal sent out via the antenna.

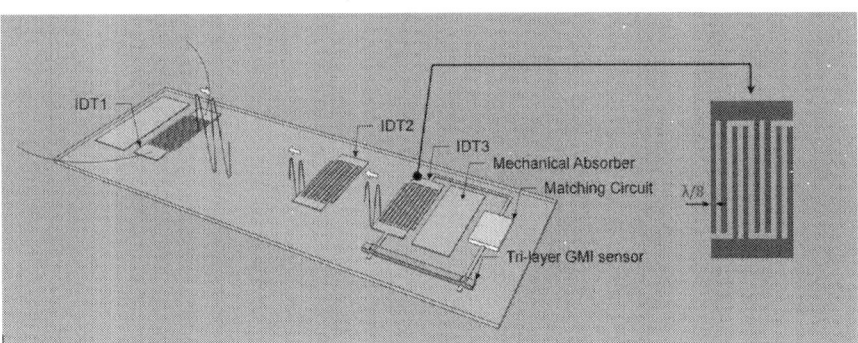

Figure 10. Schematic of an integrated passive and remote magnetic field sensor consisting of a SAW transponder and thin film GMI sensor.

In order to obtain high magnetic field sensitivity, the GMI sensor is matched to the output port (IDT3) at the working frequency of the SAW device. As the impedance of the GMI sensor changes with an applied magnetic field, the matching deteriorates, which causes the amplitude of the signal reflected from IDT3 to change. Since the piezoelectric material is sensitive to environmental changes, e.g. temperature, a reference IDT is used to provide a signal that enables the suppression of such noise by means of signal processing. Two metallic pads next to the input and output IDTs act as mechanical absorbers and suppress reflections from other structures on the substrate or the edge of the substrate.

Matching the sensor load to the optimal working point of IDT3 is a crucial aspect in the device design.Therefore, the influence of the load on the signal reflected from IDT3 issimulated. The interaction of a SAW with an IDT can be described by the P-matrix model introduced by Tobolka [63]. As shown in Fig. 11, P_{11} is the acoustic wave reflection at the output IDT [24]. Specifically, the dependence of P_{11} on the load impedance $Z = Z(Hext) + Z_m$, where $Z(Hext)$ is the impedance of the GMI element and Zm is the matching impedance, is expressed as

$$P_{11}(Z) = P_{11,sc} + \frac{2 \cdot P_{13}^2}{P_{33} + \frac{1}{Z}},$$

(17)

where$P_{11,sc}$ is the short circuit reflection coefficient, P_{13} is the electro-acoustic transfer coefficient andP_{33} is the input admittance of the transducer. In order to have a large change of P_{11}, which is equivalent to the sensitivity of the SAW device loaded by an impedance sensor, the influence of Z inequation (17) needs to be large. Therefore, a SAW transducer with a small $P_{11,sc}$and largeP_{13} will provide a large sensitivity. $P_{11,sc}$ can be minimized by using a double electrode IDT design as shown in Fig. 10, which provides cancelation of the internal mechanical reflections of the IDT.

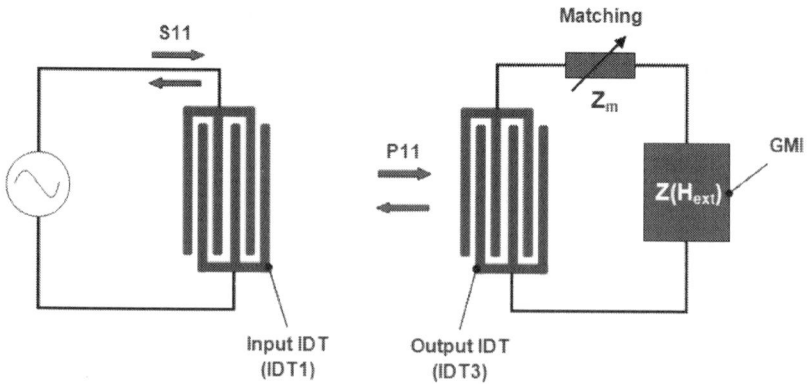

Figure 11. Electric and acoustic ports of the SAW sensor

The electro-acoustic transfer coefficient P_{13} and input admittance P_{33} can be obtained by,

$$P_{13} = \frac{1}{2\sqrt{2Z_a}} r_m (1 - e^{-j\varphi_m})$$

(18)

$$P_{33} = j\omega C_{IDT} + \frac{r_m^2}{Z_a}(1 - e^{-j\varphi_m})$$

(19)

Where r_m is the ratio of the electrical to acoustical transformer, C_{IDT} is the capacitance of the IDT, Z_a is the acoustic impedance and φ_m is the transit angle [63].

Since the GMI sensor is an inductive element, matching is accomplished by a series capacitance resulting a load impedance

$$Z = 1/j\omega C_m + R + j\omega L(H_{ext}),$$

(20)

where C_m is the matching capacitance, R is the average resistance (over the considered magnetic field range) of the GMI sensor and $L(H_{ext})$ the inductance of the GMI sensor.

Fig. 12 (a) shows the simulation result of the IDT's reflectivity as a function of the load. The slope of this plot corresponds to the magnetic field sensitivity. Therefore, the optimum matching capacitance can be determined. Fig. 12 (b) presents the rate of change of P_{11} for 1nH inductance changes (corresponding to a field change of approximately 50A/m). The result shows that with the optimum matching capacitance a maximum reflectivity change rate of 0.3dB/nH can be achieved. As the fabricated GMI sensor has an inductance change from 5nH to 15nH, a reflectivity change of 3dB can be expected.

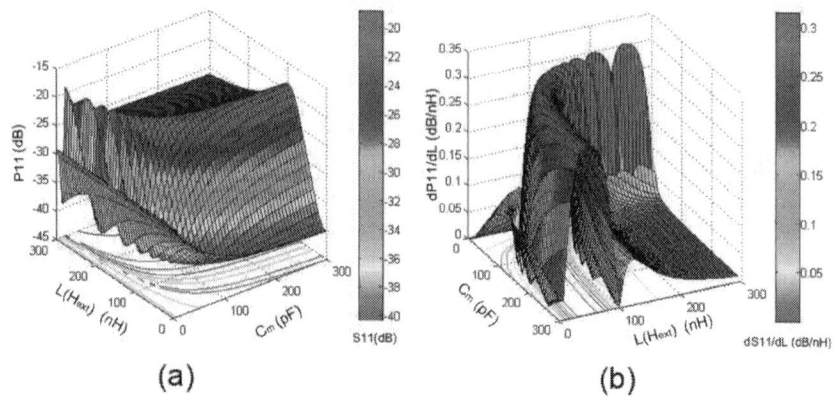

(a)   **(b)**

Figure 12. a) Magnitude P_{11} as a function of the matching capacitance and sensor inductance. (b) Rate of change (absolute value) of P_{11} for 1nH load inductance change.

The piezoelectric substrate chosen for this application is LiNbO$_3$ as it provides a strong electromechanical coupling corresponding to a high value of P_{13}. The detailed design parameters of the SAW transponder are shown in Tab. 3. The working frequency of the device is 80MHz, resulting in a periodicity p of 50µm (Equ. (1). The value of p determines the electrode width and gap. The distances between the IDTs yield a 1.25µs delay between IDT1 and IDT2 and a 0.625µs delay between IDT2 and IDT3.

Table 3 Design parameters for the SAW device.

Design parameter		Design parameter	
Substrate material	LiNbO$_3$ (128 deg. Y-X cut)	Electrode material	Gold
Center frequency	80MHz	Aperture	30λ
Periodicity	50µm	Electrode/gap width	6.25µm
Electrodes per segment	4	IDT segment number	30

The GMI sensor consists of a tri-layer structure with two ferromagnetic layers of 100nm in thickness made of Ni$_{80}$Fe$_{20}$ and a conducting copper layer with a thickness of 200nm. The sensor has a rectangular geometry of 100 µm× 4000 µm. The conducting layer is connected to the IDT3 [18].

Fabrication

The fabrication of the combined device is accomplished in several steps as shown in Fig. 13. On a $LiNbO_3$ wafer, a 40 nm Ti adhesion layer and 200 nm gold layerare sputter deposited and patterned by ion milling into individual SAW devices. The leads and SMD footprints are designed together with the SAW device to facilitate an on-chip impedance matching circuit, which was accomplished by a 150pF capacitor connected in series with the GMI element.the GMI element comprises a tri-layer structure ($Ni_{80}Fe_{20}(100nm)/Cu(200nm)/Ni_{80}Fe_{20}(100nm)$) deposited at room temperature with a uniaxial magnetic field of 200 Oe applied in the transversal direction.

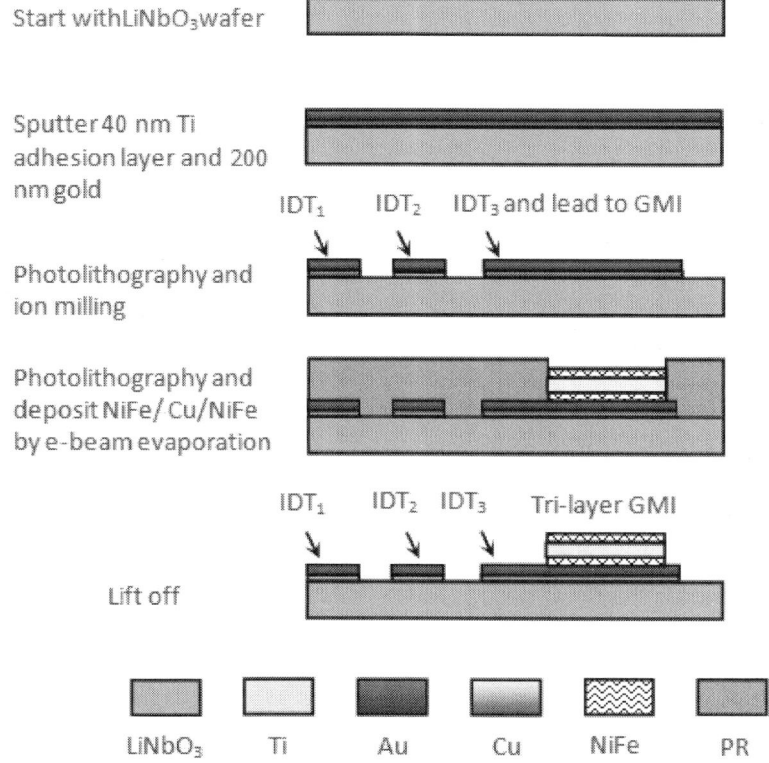

Figure 13. Fabrication flow chart of the integrated SAW-GMI device.

Results

A network analyzer (Agilent E8363C) is used to applyan RF signal to IDT1 and measure the electric reflection coefficient (S_{11}) of the input IDT, which is related to the admittance matrix of the whole device and P_{11}.The time domain signal of S_{11} is converted from the frequency domain using fast Fourier transform. As shown in Fig. 14 (a),two reflection peaks at 2.45μs and 3.55μs are observed indicating the reflections from the reference IDT and the load IDT accordingly. The magnetic response of the integrated device is determined by applying a variable magnetic field in longitudinal direction to the device. A 2.4dB amplitude change of the reflection signal can be observed. A comparison of the simulated and experimental results together with the measured GMI ratio curve is shown in Fig. 14 (b).

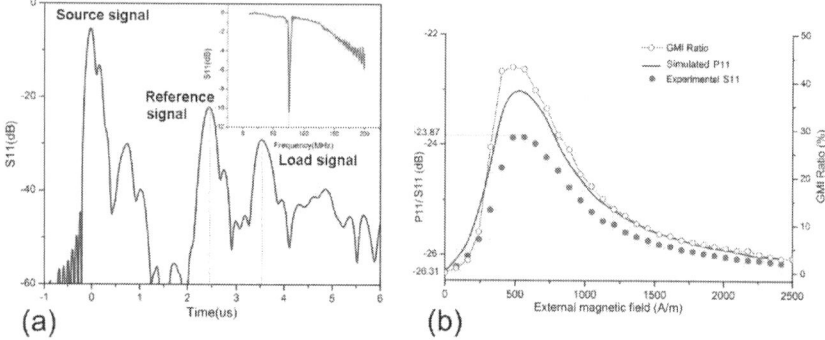

Figure 14. a) Time domain measurement of the SAW-GMI device. Inset: Frequency domain measurement. (b) Comparison of the simulated and experimental device response together with the measured GMI ratio curve.

POTENTIAL APPLICATIONS

Magnetic field sensors are one of the most widely used sensors and employed for many different applications. Current commercial magnetic sensors are wire connected to a circuit providing power and readout. These wire connections prevent the sensors from being used for certain applications. In addition, as the complexity and the number of devices, increases in modern systems such as automobiles, wire connections are becoming an increasing problem due to limited space. For those and other reasons, wireless solutions are being much sought after.

As pointed out in the previous sections, SAW-based sensors have been developed for different applications, and this technology also provides a platform for realizing wireless and passive magnetic sensors. They can provide solutions for various applications, for example, where the sensors have towithstand harsh environmental conditions or are required to have a long lifetime without maintenance.

Out of the countless applications for SAW-based passive and remote magnetic field sensors, a few will be highlighted in the following.

Nanotechnology and miniaturized systems are becoming increasingly popular in the biomedical field. Technologies based on magnetic effects are of particular interest since they can be controlled remotely via magnetic fields. For example, NVE Corporation recentlydeveloped a battery operated magnetic sensor to be used as a magnetic switch for implantable devices. When a magnetic field is applied, the sensor turns on triggering a specified action. It turns off when the field is removed. The sensor works at a stable operating point of 15 Oe [64]. Magnetic beads have been extensively used in many biomedical applications. These magnetic beads are made of ferromagnetic material ranging in size between 5 nm to 500 um. A new application of such particles promises benefits in cancer therapy by employing the particles either as drug carriers or heat sources (hyperthermia) [65]. In order to have better control of the treatment, magnetic sensors are considered to measure and detect the concentration of these magnetic particles.

The automotive industry extensively uses magnetic sensors for different purposes, for example, to measure current in electric vehicles [66] or the rotation speed of gears [67]. Another application employs magnetic sensors to detect passing vehicles using lane markers [68]. Such a system could also be used to detect vehicle speed by measuring the time between two markers of a fixed distance. In yet another application, developed by Stendl et al [69], the wear of a vehicle's tire is detected by measuring the field of magnetic beads embedded in the rubber of tire treads. As the tread size decreases, the magnetic field alsodecreases. A wireless magnetic sensor is placed just below the threads.

Construction monitoring is an upcoming application for wireless sensor. Long-term monitoring of metallic reinforcements in, e.g., bridges or buildings requires passive and remote sensors, which are capable of detecting changes of the metal. Similarly, the detection of internal defects or corrosion of pipelines is of great interest. Gloria et al [70] developed an Internal Corrosion Sensor (ICS) consisting of a magnet and a

Hall sensor. A disturbance in the magnetic field caused by changes of the metal changes the sensor readout. This information is used for both to detect and size the defects.

CONCLUSION

In this chapter we discussed different types of SAW-based, magnetic sensors including resonators, phase shifters and loaded transponders. Sensitivity to magnetic fields can be achieved byeither changing the properties of the IDT or delay line utilizing magnetostrictive materials or loading the output IDT with a magnetic field sensor. GMI sensors feature a very high sensitivity and wide range of operating frequencies and, therefore, constitute an especially suitable load. The principle of GMI sensors is described in detail and different GMI structures are discussed. While the highest sensitivity has been obtained with GMI microwires, thin film GMI sensor are advantageous because they can be produced using standard microfabricationmethods, and they can be easily integrated with a SAW transponder on the same substrate. These features are crucial with respect to production complexity and costs.

A SAW transponder combined with a GMI element connected to the output IDT is a passive and remote magnetic field sensor, which responds to an interrogation signal with a delayed response signal. The design of such a device needs to take into account different aspects like operation frequency, dimensions of IDTs and delay line or matching the load with the output IDT. In order to obtain a high sensitivity, an impedance change of the GMI element caused by a magnetic field, has to yield a large change in the SAW reflected from the output IDT. A model is presented to simulate the electro-acoustic interaction of the output IDT with the GMI sensor's impedance and the impedance matching capacitance. The simulation results provide information regarding the matching parameters and are invaluable for obtaining anoptimized performance. A detailed description of the fabrication of an integrated SAW-GMI sensor is providedusingstandard microfabricationtechnologies.The GMI ratio of the fabricated sensor is 45 % The SAW-GMI sensor provides a sensitivity of 3 dB/mT, and its output corresponds well with the simulation results.

Magnetic field sensors have countless applications and are widely used in many different fields. The trend towards wireless operation, which is generally observed nowadays, drives the development of passive and remote magnetic field sensors. Several concepts of such sensors employing SAW devices have been presented in this chapter. The most promising one is a SAW-GMI sensor, which has been discussed in detail and which features wireless and battery less operation as well as durability and the ability to withstand harsh environments. This kind of sensor is considerable not only for providing existing applications with a wireless mode; it also largely expands the potential applications of magnetic field sensors

REFERENCES

1. H. Stockman, 1948Communication by means of reflected powerProceedings of IRE. 3611961204
2. F. Plath, O. Schmeckebier, M. Rusko, T. Vandahl, H. Luck, F. Moller, D. C. Malocha, 1994Remote sensor system using passive SAW sensors. ULTRASON. 1585588
3. L Reindl, , G Scholl, , Ostertag, , Scherr H, Wolff U,Schmidt F (1998) Theory and application of passive SAW radio transponders as sensors. IEEE Trans. Ultrason., Ferroelectr.,Freq. Control. 45: 1281-1292
4. Reeder, and CullenD E (1976) Surface-acoustic-wave pressure and temperature sensors. Proceedings of the IEEE. 64:754-756
5. K. Lee, W. Wang, G. Kim, and S. Yang, 2006Surface Acoustic Wave Based Pressure Sensor with Ground Shielding over Cavity on 41° YX LiNbO3.Jpn. J. Appl. Phys. 4559745980
6. A. Pohl, 1997Wirelesslyinterrogable surface acoustic wave sensors for vehicular applications. IEEE T Instrum Meas. 4610311038
7. G S. Calabrese, H. Wohltjen, M K. Roy, 1987Surface acoustic wave devices as chemical sensors in liquids: Evidence disputing the importance of Rayleigh wave propagation, Anal. Chem. 59833837
8. Kadota, , Ito S, Ito Y, Hada T, And Okaguchi K (2011) Magnetic Sensor Based on Surface Acoustic Wave Resonators.Jpn. J. Appl. Phys. 50: 07HD07
9. S M. Hanna, 1987Magnetic Field Sensors Based on SAW Propagation in Magnetic Films.IEEE Trans. Ultrason., Ferroelectr.,Freq. Control. UFFC-34191194
10. http://wwwasrdcorp.com
11. http://wwwsenseor.com
12. http://wwwtransense.co.uk
13. Ganguly, , DavisKL, Webb DC,Vittoria C, Forester DW (1975) Magnetically tuned surface-acoustic-wave phase shifter. Electronics Letters. 11: 610-611
14. H. Hauser, R. Steindl, HausleitnerCh, PohlA,NicolicsJ (2000Wirelessly Interrogable Magnetic Field Sensor Utilizing Giant Magneto-Impedance Effect and Surface Acoustic Wave DevicesIEEE T INSTRUM MEAS. 49648652

15. Steindl, , HausleitnerCh, PohlA, HauserH, Nicolics J (2000) Passive wirelessly requestable sensors for magnetic field measurements.sens Actuators A. 85: 169-174

16. H Hauser, , J Steurer, , Nicolics, , Musiejovsky L, Giouroudi I (2006) Wireless Magnetic Field Sensor. Journal of Electrical Engineering57: 9-14

17. Al RowaisHLi B, Liang C, Green S, Gianchandani Y, Kosel J (2011Development of a Passive and Remote Magnetic Microsensor with Thin-Film Giant Magnetoimpedance Element and Surface Acoustic Wave TransponderJ. Appl. Phys. 109: 07E524

18. Li, , MH. Salem N P, Giouroudi I,Kosel J (2011) Integration of Thin Film Giant Magneto Impedance Sensor and Surface Acoustic Wave Transponder.J. Appl. Phys.111: 07E514

19. B K. Sinha, H F. Tiersten, 1979Zero temperature coefficient of delay for surface waves in quartz. Appl. Phys. Lett. 34: 817

20. Y. Ebata, H. Suzuki, S. Matsumura, and K F. Toshiba, 1982SAW Propagation Characteristics on Li2B4O7.Jpn. J. Appl. Phys. 22160162

21. Tsubouchi, .Sugai K,Mikoshiba N (1982) Zero Temperature Coefficient Surface-Acoustic-Wave Devices Using Epitaxial AlN Films. Ultrasonics Symposium, 340- 345

22. N. Dewan, M. Tomar, V. Gupta, and K. Sreenivas, 2005Temperature stable LiNbO3 surface acoustic wave device with diode sputtered amorphous TeO2 over-layer. Appl. PhysLett. 86: 223508 EOF

23. M. Viens, J. Cheeke, D N (1990Highly sensitive temperaturesensor using SAW resonator oscillator. Sensors and Actuators A: Physical 24209211

24. W. Wang, K. Lee, I. Woo, I. Park, S. Yang, 2007Optimal design on SAW sensor for wireless pressure measurement based on reflective delay lineSensors and Actuators A: Physical. 13926

25. Schimetta, ,Dollinger F, Scholl G, WeigelR (2001) Optimized design and fabrication of a wireless pressure and temperature sensor unit based on SAW transponder technology. Microwave Symposium Digest, 2001 IEEE MTT-S International. 1: 355-358

26. C. Lim, W. Wang, S. Yang, K. Lee, 2011Development of SAW-based multi-gas sensor for simultaneous detection of CO2 and 2Sensors and ActuatorsB: Chemical. 154916

27. L. Reindl, C. Ruppel, C W. Kirmayr, A. Stockhausen, N. Hilhorst, M A. Balendonck, J. Radio-requestable, Passive SAW Water-Content Sensor, IEEE Transactions on Microwave Theory And Techniques. 49: 803

28. S Q. Wang, J. Harada, and S. Uda, 2003A Wireless Surface Acoustic Wave Temperature Sensor Using Langasite as Substrate Material for High-Temperature ApplicationsJpn. J. Appl. Phys. 42: 6124 EOF6127 EOF

29. Wu, (2006) Fabrication of Surface Acoustic Wave Sensors for Early Cancer Detection, Electrical Engineering, University of California, Los Angeles

30. M. Knobel, M. Vazquez, and L. Kraus, Buschow ed. K.H.J. (2003Giant Magnetoimpedance, Handbook of magnetic materials. 15497563

31. A S. Sezen, S. Sivaramakrishnan, S. Hur, R. Rajamani, W. Robbins, and B J. Nelson, 2005Passive Wireless MEMS Microphones for Biomedical Applications, J. Biomech. Eng. 127: 1030
32. U. Wolff, F. Schmidt, G. Scholl, V. Magori, 1996Radio accessible SAW sensors for non-contact measurement of torque and temperatureUltrasonics SymposiumProceedings. 1359362
33. Yoshizawa, , Yamamoto I, and Shimada Y (2005) Magnetic Field Sensing by anElectrostrictive/MagnetostrictiveComposite Resonator. IEEE TRANSACTIONS ON MAGNETICS41: 11
34. S. Dong, J. Zhai, J. Li, and D. Viehland, 2006Small Dc Magnetic Field Response Of Magnetoelectric Laminate CompositesApplied Physics Letters82907 EOF
35. V. Koeninger, Y. Matsumura, H H. Uchida, H. Uchida, 1994Surface acoustic waves on thin films of giant magnetostrictive alloysJ ALLOY COMPD 211/212581584
36. H Uchida, , M Wada, , K Koike, , H H Uchida, ,V Koeninger, , Y Matsumura, , Kaneko, ,Kurino T (1994)Giant magnetostrictive materials: thin film formation and application to magnetic surface acoustic wave devices. J ALLOY COMPD211/212: 576-580
37. Panina, andMohri K (1994) Magneto-impedance effect in amorphous wires.Appl.Phys.Lett. 65: 1189-1191
38. Machado, L A, RezendeS M (1996)A theoretical model for the giant magnetoimpedance in ribbons of amorphous soft-ferromagnetic alloys. Journal of Applied Physics. 79: 6558- 6560
39. Hika, ,Panina LV,Mohri K (1996) Magneto-Impedance in Sandwich Film for MagneticSensor Heads. IEEE Trans Magn. 32: 4594-4596.
40. Panina, ,MakhnovskiyD P, MappsD J, and ZarechnyukD S (2001) Two-dimensional analysis of magnetoimpedance in magnetic/metallic multilayers. J. Appl. Phys. 89: 7221
41. T. Morikawa, Y. Nishibe, H. Yamadera, Y. Nonomura, M. Takeuchi, Y. Taga, 1997Giantmagneto-Impedance Effect in Layered Thin Films. IEEE Trans Magn. 3343674372
42. A. De Andrade, M H. Da, R B. Silva, M A. Correa, A. Viegas, D C. Severino, A M. Sommer, R L (2004Magnetoimpedance of NiFe/Ag multilayers in the 100 kHz-1.8 GHz range. Journal of Magnetism and Magnetic Materials 272-276: 1846-1847
43. V. Zhukova, A. Chizhik, A. Zhukov, A. Torcunov, V. Larin, and J. Gonzalez, 2002Optimization of giant magnetoimpedance in Co-rich amorphous microwires.IEEETrans. Magn. 3830903092
44. A T Le, , M H Phan, , C O Kim, , M Vázquez, , H Lee, , Hoa, andYu S C (2007) Influences of annealing and wire geometryon the giant magnetoimpedance effect in aglass-coated microwire LC-resonator, J. Phys. D: Appl. Phys. 40: 4582-4585

45. M Vázquez, , Adenot-Engelvin, (2009) Glass-coated amorphous ferromagnetic microwires at microwave frequencies. Journal of Magnetism and Magnetic Materials. 321: 2066-2073

46. Z. Zhou, Y. Zhou, L. Chen, 2008Perpendicular GMI Effect in Meander NiFe and NiFe/Cu/NiFe FilmIEEE Transactions on Magnetics4422522254

47. F Pompéia, , Gusmão, A P, Hall BarbosaC R, Costa MonteiroE, Gonçalves L A P and Machado F L A (2008) Ring shaped magnetic field transducer based on the GMI effect. Meas. Sci. Technol. 19: 025801

48. M H. Phan, H X. Peng, 2008Giant magnetoimpedance materials:Fundamentals and applications.Progress in Materials Science. 53323420

49. P. T. Squire, D. Atkinson, Gibbs MRJ, Atalay SJ (1994Amorphous wires and their applicationsJ MagnMagnMater. 1321021

50. I. Ohnaka, T. Fukusako, T. Matui, 1981Preparation of amorphous wires. J.Jpn. Inst. Met. 4575162

51. http://wwwaichi-steel.co.jp/ENGLISH/pro_info/pro_intro/elect_3.html

52. L. V. Panina, K. Mohri, T. Uchiyama, M. Noda, 1995Giant magneto-impedance in Co-rich amorphous wires andfilms. IEEE Trans Magn. 31124960

53. M. H. Phan, H. X. Peng, S. C. Yu, M. Vazquez, 2006Optimized giant magnetoimpedance effect in amorphous andnanocrystalline materials. J Appl Phys. 99:08C505

54. A L. Sukstanskii, and V. Korenivski, 2001Impedance and surface impedance of ferromagnetic multilayers: the role of exchange interaction J. Phys. D 34: 3337

55. B. Li, J. Kosel, 2001d Simulation of GMI Effect In Thin Film Based Sensors. J. Appl. Phys. 109: 07E519

56. C. Dong, S. Chen, Hsu T Y (XuZuyao) (2003A modified model of GMI effect in amorphous films with transverse magnetic anisotropyJ. Magn. Magn. Mater. 2637882

57. S. Xiao, Y. Liu, Y. Dai, L. Zhang, S. Zhou, and G. Liu, 1999Giant MagnetoimpedanceEffect in Sandwiched Films. J. Appl. Phys. 85: 4127

58. S. Xiao, Y. Liu, S. Yan, Y. Dai, L. Zhang, and L. Mei, 2000Giant Magnetoimpedance and Domain Structure in FeCuNbSiBFilms and Sandwiched Films. Phys. Rev. B 6157345739

59. D. De Cos, L. V. Panina, N. Fry, I. Orue, A. Garcia-arribas, J. M. Barandiaran, 2005Magnetoimpedance in narrow NiFe/Au/NiFe multilayer film systemsIEEE Transactions onMagnetic. 4136973699

60. Li, , YuanW, ZhaoZ, Ruan J and Yang X(2005) The GMI effect in NanocrystallineFeCuNbSiBMultilayeredFilms with aSiO2 Outer Layer. J. Phys. D: Appl. Phys. 38: 1351-1354

61. M A. Corrêa, F. Bohn, V M. Escobar, M S. Marques, A. Viegas, D C. Schelp, L F. Sommer, R L (2011Wide Frequency Range Magnetoimpedance in Tri-layered Thin NiFe/Ag/NiFeFilms: Experiment and Numerical Calculation. J. Appl. Phys. 110: 093914

62. Z. Zhou, Y. Zhou, L. Chen, and C. Lei, 2011Transverse, Longitudinal and Perpendicular Giant Magnetoimpedance Effects in a Compact MultiturnMeanderNiFe/Cu/NiFeTrilayerFilm Sensor. Meas. Sci. Technol. 22: 035202

63. G. Tobolka, 1979Mixed matrix representation of SAW transducers. IEEE Trans. Sonics Ultrason. SU-26

64. http://wwwmedicalelectronicsdesign.com/products/nanopower-magnetic-sensors-fit-implantable-devices

65. J L. Corchero, A. Villaverde, 2009Biomedical Applications of Distally Controlled Magnetic Nanoparticles.Trends in Biotechnology27468476

66. P. Ripka, 2008Sensors based on bulk soft magnetic materials: Advances and challengesJournal of Magnetism and Magnetic Materials32024662473

67. J. Lenz, S. Edelstein, 2006Magnetic sensors and their applications. IEEE Sensors Journal. 6631649

68. Nishibe, ,Ohta N,Tsukada K,Yamadera H,Nonomura Y,Mohri K, Uchiyama T (2004) Sensing of passing vehicles using a lane marker on a road with built-in thin-film MI sensor and power source. IEEE Transactions on Vehicular Technology. 53: 1827- 1834

69. R. Steindl, C. Hausleitner, H. Hauser, W. Bulst, 2000Wireless magnetic field sensor employing SAW-transponder," Applications of Ferroelectrics. Proceedings of the 2000 12th IEEE International Symposium on. 2855858

70. N. Gloria, B S. Areiza, M C L. Miranda, I V J. Rebello, J M A (2009Development Of A Magnetic Sensor for Detection And Sizing of Internal Pipeline Corrosion DefectsNDT & E International. 42669677

CITATION

Bodong Li, Hommood Al Rowais and Jürgen Kosel (2013). Surface Acoustic Wave Based Magnetic Sensors, Modeling and Measurement Methods for Acoustic Waves and for Acoustic Microdevices, Prof. Marco G. Beghi (Ed.), ISBN: 978-953-51-1189-4, InTech, DOI: 10.5772/55220.

Chapter 4

Auditory Perceptual Objects as Generative Models: Setting the Stage for Communication by Sound

István Winkler [1,2,] Erich Schröger [3.]

[1] Institute of Cognitive Neuroscience and Psychology, Research Centre for Natural Sciences, Hungarian Academy of Sciences, Hungary
[2] Institute of Psychology, University of Szeged, Hungary
[3] Institute for Psychology, University of Leipzig, Germany

ABSTRACT

Communication by sounds requires that the communication channels (i.e. speech/speakers and other sound sources) had been established. This allows to separate concurrently active sound sources, to track their identity, to assess the type of message arriving from them, and to decide whether and when to react (e.g., reply to the message). We propose that these functions rely on a common generative model of the auditory environment. This model predicts upcoming sounds on the basis of representations describing temporal/sequential regularities. Predictions help to identify the continuation of the previously discovered sound sources to detect the emergence of new sources as well as changes in the behavior of the known ones. It produces auditory event representations which provide a full sensory description of the sounds, including their relation to the auditory context and the current goals of the organism. Event representations can be consciously perceived and serve as objects in various cognitive operations.

COMMUNICATION CHANNELS

Communication requires a channel open between the participants allowing them to exchange information. Communication by sound typically occurs in environments rich in sound sources. In order to listen to someone speaking, we have to be able to create and maintain the channel conveying the information provided by the speaker. This involves separating the speaker's voice from all concurrent streams of sound which themselves are potential alternative channels to choose. For example, while driving a car, we can hear the sound of the car engine, the noise of the tires rolling over the surface, music from the radio while still being able to conduct a conversation with another person. Parsing the mixture of sounds arriving at our ears (termed Auditory Scene Analysis; Bregman, 1990) results in the formation of perceptual units called auditory objects (e.g. the speaker's voice; Griffiths and Warren, 2004, Kubovy and van Valkenburg, 2001 and Winkler et al., 2009).

Every-day experience tells us that sounds deviating from the acoustic context often break into our conscious experience even if previously we did not attend their source. For example, in the previous mentioned situation (i.e., having a conversation while driving a car), one typically only notices the sound of the car engine, if it starts to cough. Deviance detection has been often studied using electric brain responses elicited by auditory events, termed auditory event-related potentials (ERPs). Sounds violating some regular feature of the preceding sequence have been shown to elicit a specific component within the auditory ERPs, termed the mismatch negativity (MMN; Näätänen, Gaillard, & Mäntysalo, 1978; for reviews, see Kujala et al., 2007 and Näätänen et al., 2011). Human and animal research in the past 30 years have revealed many details about how auditory scenes are analyzed, as well as how deviant sounds are detected within the auditory system. However, the two areas of research – auditory scene analysis and auditory deviance detection – have proceeded largely independently from each other. Here, we provide an integrative research review that develops connections between these two areas.

One common thread between the two functions is that they both require some representation of the immediate history of the stimulation. Such a representation allows discrete sounds to be linked together to form an auditory perceptual object, as well as to assess whether they carry new information with respect to what we already know about the sound sources in the environment. We will argue that a second common feature is that both auditory scene analysis and auditory deviance detection look into the future. That is, we provide a theoretical framework linking auditory scene analysis and deviance detection via predictive auditory representations.[1]

The idea of human information processing and specifically perception operating in a predictive manner has a long tradition both in psychology and neuroscience. For example, Gregory's (1980) influential contemporary empiricist theory likens perception to scientific hypotheses, which provide the brain's "best guess" of the causes (distal objects) of the stimulation reaching the sensory organs (the proximal stimuli) and can produce extrapolations to parts of the environment, which are currently not accessible to the senses. Recent theories following the empiricist tradition, which started with Helmholtz's (1867) notion of unconscious inference and has been arguably the most influential school for explaining perception (see, e.g., Clark, 2013), posit predictive models integrating perception, attention, learning, and even actions (e.g., Ahissar and Hochstein, 2004, Bar, 2007, Friston, 2010, Hohwy, 2007, Hommel et al., 2001,Summerfield and Egner, 2009 and Tishby and Polani, 2011). In neuroscience, Helmholtz's theory coupled with Bayesian rules for optimal inference generation (Kersten et al., 2004 and Knill and Pouget, 2004) engendered the predictive coding theories appearing first in the 1990s (e.g., Mumford, 1992 and Rao and Ballard, 1999). Modern versions of predictive coding assume the existence of a hierarchy of generative models with increasing levels of abstraction (see e.g., the free energy principle of Friston, 2005 and Friston, 2010). At each level of the hierarchy, predictions from a generative model are compared with the input and the difference is treated as an error signal. The system aims at suppressing (minimizing) the error by adjusting models, with higher levels governing model selection at lower levels.

Effects of stimulus predictability have been shown on auditory scene analysis (e.g.,Andreou et al., 2011, Bendixen et al., 2010 and Rimmele et al., 2012; initially suggested by Jones, 1976; for a review, see Bendixen, 2014). Regular (predictable) tone patterns embedded separately within two interleaved sequences increased the probability of hearing two concurrent sound streams as opposed to a single streams (Bendixen et al., 2010, Bendixen et al., 2013 and Szalárdy et al., 2014), while predictable patterns connecting tones across the two interleaved sequences that did not at the same time produce such patterns separately for the two sequences increased the probability of perceiving a single stream over two concurrent ones (Bendixen, Denham, & Winkler, 2014). Further, a predictable pattern (a tune) embedded in one of two interleaved sound sequences made it easier for listeners to follow the other sound sequence (Andreou et al., 2011 and Rimmele et al., 2012). Predictive processes probably also play a crucial role in auditory deviance detection (e.g., Bendixen et al., 2012, Lieder et al., 2013 and Paavilainen et al., 2007; initially suggested by Winkler, Karmos, & Näätänen, 1996; for a review, see Bendixen, SanMiguel, & Schröger, 2012). Winkler, Karmos, et al. (1996; see also Winkler, 2007) have suggested that deviance is established by comparing incoming sounds against those predicted by the representations of previously detected regularities. For example, when a tone sequence followed the rule "long tones are followed by high ones, whereas short tones by low ones", rare low tones following long ones and high tones following short ones elicited the MMN response signaling that the rule violation was detected (Paavilainen et al., 2007; see also Bendixen, Prinz, Horváth, Trujillo-Barreto, & Schröger, 2008). In this sequence, deviant tones did not contain any rare feature of feature combination, *per se*. Only because the previous tone predicted a different tone to arrive next in the sequence made these tones to violate the acoustic regularity of the sequence, and therefore to be processed as deviants.Bendixen, Schröger, and Winkler (2009) have also found that differences between ERPs elicited by the occasional omission of a predictable vs. an unpredictable tone. These and other evidence reviewed by Bendixen, SanMiguel, et al. (2012) strongly support the notion of the involvement of predictive processes in MMN generation.

Our theoretical framework linking auditory scene analysis and deviance detection is compatible with the general idea of predictive coding. We will argue that regularities detected from the relationship between successive sounds are encoded into generative models of the acoustic environment. Predictions from these models help to construct auditory sensory memory representations and they are compared to the currently dominant interpretation of the auditory input. The outcome of the comparison is used to update the model.

Research on speech processing usually focuses on how the brain decodes spoken messages. The input of most of these models is a stream of speech. That is, they assume that the communication channel is already established. Here we provide a conceptual framework for how the auditory system sets the stage for this. Since using predictions to reduce the amount of computation required to decode messages have also been suggested for language processing (Federmeier, 2007, Hosemann et al., 2013 and van Petten and Luka, 2012), the model proposed here fits seamlessly with such models, specifying some lower levels of the hierarchy.

THE BUILDING BRICKS: REGULARITY, DEVIANCE, PREDICTIVE INFORMATION PROCESSING

Deviance can only be defined in relation to something regular. An event is deviant if it does not fit at least one of the relationships connecting the previous events within the environment. That is, a deviant event violates some existing regularity of the context within which it appears. By regularity we mean an implicit sequential rule, which is extracted from the series of sound events by the auditory system. Later, we will specify the types of regularities involved in auditory deviance detection (e.g., concrete and statistical regularities), how they are utilized, and how such regularities are extracted from a sequence of sound. In the auditory modality, deviations range from simple cases, such as breaking the repetition of a discrete sound, to complex ones, such as violating a harmonic or rhythmic rule in music. From the above definition follows that within a sequence of sounds with no regular relationships no sound event can be deviant. Another consequence is that deviance is not equal to physical (acoustic) change. Let us consider a spoken sentence with monotonously falling pitch (such as is typical in statements spoken in Hungarian). Although the pitch of each word is different from the previous one, because it fits the regularity, it is not a pitch deviant. On the other hand, while a word having the same pitch as the previous one represents no pitch change it deviates from the pitch regularity of the preceding ones (i.e., it is a pitch deviant event).[2]

When describing perception, the above definition of deviance should be further specified in order to take into account the capabilities of the perceiver, that is, the system that would detect deviance. Because the presence of regularity is a prerequisite of deviance detection, the system can only detect deviants that break some regularity that the system "knows about". One can only detect a rhythmic violation in a poem if one remembers the poem or if the word violates some general rhythmic convention the person has experience with. This means that for detecting deviance, the system (in our case the human auditory system) must have access to some representation of the regular relationships applicable to the current environment as well as mechanisms which allow it to determine whether or not a given sound matches these regularity representations.

One should also consider the environment. For the perceiver, the environment is not equal to the physical effects reaching the senses. Our experience (stored representations) of the environment co-determines what we detect as deviant. In our previous example, someone regularly listening to poems in English would detect violations of meter even in English poems he/she never heard before. Thus when we refer to the environment, we mean the combination of two things: the environment and the listeners pre-existing representations of this environment, the context.

What is common between an acoustic regularity established by the recently encountered sounds (such as a sequence of two alternating tones) and rhythmic conventions in poems? They both allow one to predict which sounds are likely to follow the ones just heard. We shall argue that in the human auditory system, regularity representations are used to generate predictions for future events and incoming sounds are checked against these predictions. Consider the situation of crossing a street: We are not only interested where cars are at the moment, but, rather, where they will be when we reach their lane. Recent accounts of perception (Bar, 2007, Enns and Lleras, 2008, Ghahramani and Wolpert, 1997, Gregory, 1980, Schubotz, 2007, Summerfield and Egner, 2009 and Winkler et al., 2009) as well as computational models of sensory processes (Friston, 2005, Friston and Kiebel, 2009 and Tishby and Polani, 2011) emphasize that information processing is directed towards the future. In the same vein, we term the set of representations of the known regularities of a given environment the (predictive/generative) model of this environment and we suggest that the auditory system maintains such a predictive model of the acoustic environment.

Why is it advantageous to establish such a model? Living organisms require information for reaching their goals and to successfully adapt their behavior to the environment. Deviance is of special importance as it represents new information that may require some response from the organism. In fact, deviance in the above defined sense is equal to new information for the organism, because sounds conforming to previously detected regularities could be predicted by the organism. Having a model of the environment allows the organism to predict a part of the input and thus prepare to take appropriate action. The larger the part of the input, that the sensory systems can predict, the fewer the information that requires detailed evaluation. As a consequence, fewer resources are needed for processing the actual sensory input. In other words, it is advantageous for the organism to invest into building a good model as the model will permit it to successfully adapt to the environment while conserving resources. The predicted part of the input does not require further processing, unless the event is actively monitored (e.g., one wishes to synchronize an action with an expected event). No information is lost by this type of filtering. At the same time, by identifying deviance, new, possibly important information gets a better chance to receive detailed processing.

Such generative models may perhaps be even more important in the auditory modality than in vision, because the acoustic environment is ephemeral; it lacks elements which can be revisited at will. An important characteristic of sound is that it unfolds in terms of temporally varying signals. Even the most elementary acoustic features, such as pitch or the direction of the sound source require the processing of sound segments of some duration. Moreover, any meaningful analysis of the acoustic environment involves connecting discontinuous segments of the incoming sound flow. How else could we understand prosodic information or tell whether or not the footsteps we hear signal that someone approaches us. Therefore, in order to establish perceptual events, the system must take into account the temporal behavior of the input. The generative model of the environment to be described here serves this purpose. We shall outline the various processes involved in establishing, maintaining, and utilizing a predictive model of the acoustic environment. The information within the model serves multiple purposes, deviance detection being only one of them. We shall argue that the model provides the basis of organizing the acoustic input into perceptual units (objects), which represent the concurrently active independent sources in the environment. That is, the model to be described here is an essential element of auditory scene analysis (Bregman, 1990).

AN OVERVIEW OF DETECTING NEW INFORMATION IN THE AUDITORY MODALITY

We regard the set of processes and memory resources involved in detecting new auditory information a system with functional module-like properties.[3] The input to this functional module is sensory data analyzed for basic auditory features. On its output, it delivers a sensory event representation, which, in addition to describing the sensory features of the incoming sound, also specifies the relation of this sound to the auditory context including an evaluation of how well it conforms to the regularities detected from the preceding sounds. Therefore, we term this functional module the Auditory Event Representation System (AERS). AERS is module-like in the sense that it can properly function based on the auditory input alone (i.e., without voluntary effort or focused attention). For this reason, deviance detection has often been described as pre-attentive in the literature. However, the notion of pre-attentiveness assumes a strict serial order between stimulus-driven and attentive processing, which most likely does not hold for auditory deviance detection (see, e.g., Haroush et al., 2010 and Sussman et al., 2002). These studies have shown that top-down effects, especially those biasing how the auditory input is structured into streams and patterns affect which regularities are extracted and, as a consequence, which sounds are detected as deviants (for a review, see Sussman, 2007). Thus deviance detection is not pre-attentive. However, many deviations are detected even when attention is not focused on the sound sequence (for a review, see Näätänen, 1990). Moreover, even when the auditory input is generally unattended, AERS relies on a memory store, which interacts with other forms of memory, including long-term memory representations (for a review, see, Näätänen, Tervaniemi, Sussman, Paavilainen, & Winkler, 2001). These studies have shown that information learned on-line (such as a difficult discrimination or the structure of short trains; see e.g.,Näätänen et al., 2001, Schröger et al., 2004 and Winkler and Cowan, 2005) or previously (such as representations of the phonemes of a language spoken by the listener; see e.g.,Näätänen et al., 1997 and Winkler et al., 1999), and even automatized processing strategies (such as those learned by musicians; see e.g., Brattico et al., 2002, van Zuijen et al., 2004 and van Zuijen et al., 2005) modulates the processing of sounds. Although the memory representations and processes of AERS are not necessarily consciously experienced (i.e., they are of implicit nature) as there are deviations registered in the brain, which do not appear in conscious perception (see Paavilainen et al., 2007,Sussman et al., 2002 and van Zuijen et al., 2006), this is a weaker form of modularity compared with that defined by Fodor (1983), because of the possibility of outside access and modification of some internal processes.

Fig. 1 shows an overview of AERS. For the sake of simplicity, let us first consider the detection of deviance within a single coherent sound sequence (a single auditory stream typically delivered by a single sound source – for a discussion of the relationship between auditory streams and sound sources, see Bregman, 1990). Later we shall consider the case of multiple sound streams (complex auditory scenes; see Section 5). Let the sequence consist mostly of sounds that conform to some regularity: For example, a sequence of sounds with a common timbre, such as would be produced by a person speaking in a neutral voice. Sounds meeting the regularity are termed "standard" sounds, whereas sounds violating the regularity are termed "deviants" (e.g., a high-pitched sound of surprise in the above example). With such a simple acoustic regularity in mind, we now describe the four major constituents of AERS.

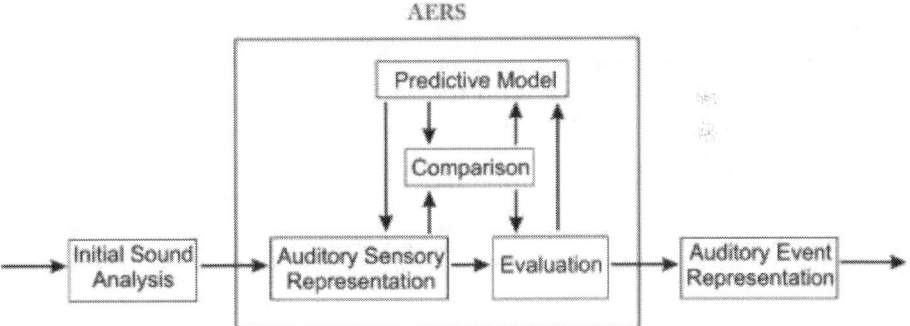

Figure 1. An overview of the auditory event representation system (AERS). The primary input to AERS is the incoming sound with its basic features established. The main components of the system include a Predictive Model of the auditory environment storing representations of regularities extracted from the preceding sounds. This model generates predictions for upcoming sounds, thus helping to establish Auditory Sensory Representations of the incoming sounds. The resulting representation is compared with the predictions. The outcome of the Comparison is used to (1) update the model and (2) evaluated together with information regarding the current goals of the organism. The result is an Auditory Event Representation (the main output of AERS), which can enter various mental operations and be consciously perceived. Also, the model is influenced by the Evaluation process, which can initiate the building of new or reactivate old but inactive regularity representations.

Forming Auditory Sensory Memory Representations

Some characteristics of the incoming sound (proto-features: such as, the possible periodicity of the signal, spectral energy maxima, binaural differences, etc., such as the speech landmarks; Stevens, 2002) are extracted early within the afferent pathways of the auditory system. The traditional view suggests that these features are then bound together to form unitary sound representations[6] (the "feature binding problem": Bertrand and Tallon-Baudry, 2000, Treisman, 1993, Treisman, 1998, Treisman and Gelade, 1980 and Zhuo and Yu, 2011). This feed-forward account is overly simplistic because even discrete sounds can be quite complex and, when two or more sounds overlap each other correctly establishing the features requires that they are first separated from each other. However, unfortunately, the literature currently provides little evidence regarding the details of the interaction between feature extraction and sound separation beyond establishing the ubiquitous presence of descending pathways throughout the auditory system (Schofield, 2010). Therefore, by necessity, we only note the probable existence of such interactions and focus on how auditory features are utilized in deviance detection and sound grouping.

Thus auditory features are bound together to form unitary auditory sensory memory representations (Fig. 1). There is good reason to suggest that the formation of these representations can be affected by the context. An example is the phonemic restoration effect, when the phoneme correctly completing a spoken word is heard even when the actual phoneme sound is omitted and the gap is filled with a sound spanning a broad frequency band (Samuel, 1981 and Shinn-Cunningham and Wang, 2008). In the phonemic restoration effect, we hear a sound that is not present as such in the acoustic input. Thus, this percept must be derived from memory, as somebody unfamiliar with the given language will not correctly restore the phoneme. The memory representation does not have to be previously learned as similar restoration effects can be observed for sounds with a predictable pitch contour (see, e.g., the continuity illusion, Riecke, van Opstal, & Formisano, 2008). This suggests that establishing unitary auditory sensory memory representations rests on the construction of predictions from memories that persist over different time scales.

There is no consensus on whether or not feature binding requires focused attention (seeTreisman, 1993, Treisman, 1998 and Treisman and Gelade, 1980; vs. Duncan & Humphreys, 1989). Based on the results of deviance detection studies (see, Näätänen, 1992, Näätänen and Winkler, 1999 and Sussman, 2007), we assume that such representations are formed even outside the focus of attention. On the other hand, we will note throughout the description of AERS how its output, termed Auditory Event Representation (Fig. 1) can be affected by attention.

Predictive Model

The model ("Predictive Model" box in Fig. 1) stores representations of the currently applicable auditory regularities (e.g., in our above example, sounds having a common timbre). The regularity representations produce predictions for upcoming classes of sound (e.g., the next sound should have the same timbre) (Baldeweg, 2006, Baldeweg, 2007, Bendixen et al., 2009, Grimm and Schröger, 2007, Winkler, 2007, Winkler et al., 1996 and Winkler et al., 2009; and also specifically to speech sounds, see Bendixen, Scharinger, Strauss, & Obleser, 2014). These predictions (a) guide the formation of auditory sensory memory representation of the incoming sound and (b) are compared with the emerging sensory memory representation of the sound (see below).

Existing regularity representations are updated when the incoming sound mismatches the predictions of the model (Schröger, 2007, Winkler, 2007 and Winkler et al., 1996) whereas new regularities are extracted once the predictable part of the auditory input is accounted for. Modulating effects on the model reflect structural information encoded in longer-term stores such as long term experience with certain types of sounds as well as explicit knowledge about the current sound sequence. The former has been shown by studies comparing "experts" and naive listeners: players of a given instrument detect smaller pitch deviations for their own instrument compared to listeners playing other instruments (Tervaniemi, Just, Koelsch, Widmann, & Schröger, 2005) and speakers of a language are superior in detecting phoneme category (Näätänen et al., 1997 and Winkler et al., 1999) and word changes (Jacobsen et al., 2004 and Pulvermüller et al., 2001) relevant in the given language than listeners, who don't speak that language. The effect of explicit knowledge on deviance detection was demonstrated in an experiment in which listeners did not notice that the sound sequence consisted of a cyclically repeating pattern until they were informed about it; their brain response to rare sounds within the pattern reflected that they structured the sequence differently afterwards (e.g., Sussman, Winkler, Houtilainen, et al., 2002).

Comparing Model Predictions with the Sensory Representation of the Current Sound

Depicted at the center of the overview of AERS is the comparison between the sensory representation of the incoming sound and the predictions derived from the model of the acoustic environment (Fig. 1). That is, we assume an explicit comparison function (cf. comparator-based deviance-detection; Opitz et al., 2005 and Siddle, 1991), although, previously implicit solutions to the comparison function have also been suggested (Näätänen, 1984). Unlike in laboratory settings, under everyday circumstances, no sound can be fully predicted. Acoustic variability is introduced by (a) variations in the sound source, (b) changes in the relative position of perceiver and the source (i.e., as they move with respect to each other) as well as by (c) various concurrent changes in the physical environment (e.g., movement of sound-reflecting objects). Therefore, (a) the models stores the distributions of feature values and its predictions are adapted to the experienced variability of the preceding sound sequence by referring to ranges of the feature space and classes of sounds, rather than to a specific sound (see, e.g., Winkler et al., 1990, who found that intensity and pitch deviations were detected despite variations in the intensity of the regular sounds); and (b) the comparison output signal (MMN) may reflect the amount of deviation (Näätänen and Alho, 1997, Schröger and Winkler, 1995 and Tiitinen et al., 1994; see, however Horváth et al., 2008), as the MMN amplitude increases and the peak latency decreases with increasing amounts of deviance (however, due to overlap between MMN and other ERP responses, it is possible that a large part of this effect originates from a different source).

The outcome of the comparison, which describes the relation between the sensory representation of the incoming sound and the regularities stored by the model, is passed on to the evaluation process (see below). If predictions from the model failed, the model needs to be corrected. Thus the outcome of the comparison feeds back to the model via an updating process. The updating process is reflected by the MMN (Winkler, 2007,Winkler and Czigler, 1998 and Winkler et al., 2009). This does not, however, rule out that the MMN signal can also serve as an indicator that new information has been encountered by the perceiver (Escera et al., 2000, Näätänen, 1990 and Schröger, 1997). Thus, the MMN can be seen as one (though not the only) indicator of new information, a brain index of prediction error, which drives the updating of the model.

Evaluation

This is the point, at which the incoming auditory information can be fully assessed and prepared for possible further processing (outside AERS) for attention control (e.g., orienting), determining the real-life event that gave rise to this sound, and assessing its relevance for the current or some pre-planned actions of the organism. Evaluation takes into account the context set up by the current goal-directed processes (top-down effects). Consider the case of timbre constancy as a detected regularity. Usually, small variation in any stimulus parameter is tolerated. If, however, someone were looking for signs of emotional stress in another person's voice, he/she would very likely notice even very small timbre deviations. Another possible reason for marking a sound for further processing is when it or its relation to the preceding auditory context meets some preset pattern (Formby, 1967 and Roye et al., 2007) such as hearing one's own name (Perrin, Garcia-Larrea, Mauguiere, & Bastuji, 1999). On the other hand, even relatively large sound deviations could go unnoticed, when one's attention is strongly focused somewhere else. Thus the evaluation of a sound takes into account those aspects of the context which are outside the auditory environment. The resulting information package is the primary output of AERS. We term this an auditory event representation, because it describes the sound together with its relation to both the auditory and the general context.

The other main function of the evaluation processes is to initiate the search for new regularities. Successive deviant events may signal a change of the sound source or its behavior. The full description of the acoustic event can be used to find new regularities within the acoustic environment. Furthermore, this is the point at which the unpredicted part of the input (the residue) can be assessed. The residue may reflect the emergence of a new sound source in the environment. For example, typically, no predictions exist for the emergence of a new voice (a new instrument or person) within the auditory input. Such sounds remain unaccounted by the active generative models, forming the residue, after all predictions are checked. Note that although both deviations from the predictions of existing regularity representations and the residue are de facto prediction errors, the former is specific to a given regularity representation, the latter is general to the whole model and they are utilized differently in AERS: updating a regularity representation vs. initiating the formation of a new one.

The P3a ERP response may reflect the assessment of the information value of the incoming sound by the evaluation process. The currently most widely accepted interpretation of P3a is that it reflects a call for further processing of highly deviant or otherwise unexpected stimulus events (e.g., a sound delivered after a long silent interval; Friedman et al., 2001 and Polich, 2007). However, this interpretation has been challenged (Horváth et al., 2008, Rinne et al., 2006 and Wetzel et al., 2013), because some features of P3a suggest that it reflects the assessment of the "significance" of sensory events, combining the information carried by the stimulus with its relevance within a wider context. The latter interpretation is compatible with the assumed function of "evaluation" within AERS.

INITIAL BUILD-UP OF AN AUDITORY MODEL

It has to be asked how a representation of an auditory regularity is established, such as when the sound sources in the environment have been inactive for some time and, therefore, previously detected regularities may not be available, or when a new regularity comes into play. We will illustrate this for simple regularities, before we consider more complex ones.

Simple Regularities

When a sound arriving after a longer silent period repeats a few times, each sound reaching the ears receives initial analysis in the afferent auditory pathway (Carney, 2002). Units of the input are typically separated by abrupt spectrotemporal changes (onsets and offsets), which may be indexed by the elicitation of the onset- and offset-related ERPs, such as the N1 response (Näätänen & Picton, 1987). Markers for the start of sound-units serve as temporal reference points, allowing the auditory system to compare sound patterns with the corresponding representations along the temporal axis, such as distinguishing the A-B tone pair from B-A tone-pair. Evidence shows that the later parts (>350 ms) of long sounds only affect the building of regularity representations when the sound includes abrupt spectrotemporal changes (Schwartze et al., 2012 and Weise et al., 2010). This suggests that representations are based on the initial segment of long continuous sounds. However, abrupt spectrotemporal changes within a long sound (such as most consonants) initiate the formation of a new unit, thus enabling a segmented, but precise description of the full sound. Our notion of the basic unit is compatible with that of the literature of auditory sensory memory (see Cowan, 1987 and Demany and Semal, 2008) as well as with the notion of parallel analysis of sound on multiple times scales within the human auditory cortex (Nelken et al., 2003 and Poeppel, 2003).

Because we assumed that no competing regularity representations preexist, the established sound features are conjoined unless they contain some spectral or temporal cue indicating the presence of multiple concurrent sound sources, such as a mistuned harmonic within a complex tone or asynchronous onset of different spectral components (for reviews of the cues supporting the instantaneous segregation of concurrent sounds, see Alain, 2007, Carlyon, 2004, Ciocca, 2008 and de Cheveigné, 2001). Note, that we postpone discussing multi-source sound configuration till later. Here we assume that a unitary representation of a single sound is formed through projecting the different auditory features onto temporal coordinates, thus constructing a description encoding both static and dynamic aspects of the discrete sound (Näätänen & Winkler, 1999). The resulting representation may then be consciously perceived.

Because we consider the (rare) case that no regularity representations were available when the sound was encountered, no predictions could have been formed for this sound. Thus the whole auditory input becomes "residue" within AERS and thus the formation of a new regularity representation is triggered. Because sounds delivered after a long silent period elicit very large responses in auditory cortex (as characterized by the P1 and N1 ERPs), some part of these auditory cortical responses may be related to the process initiating the formation of a new regularity representation (Winkler et al., 2009). One possibility is that the search for a new regularity is initiated when the strength of these responses exceeds some threshold (for a similar idea referring to memory traces, seeNäätänen, 1984). However, a single sound is not sufficient for establishing a regularity representation. This has been demonstrated by the lack of any deviance-related ERP response when the second sound of a train differed from the first one while the train was preceded by a long silent period and no compatible regularity had been established in the preceding train (Cowan et al., 1993 and Winkler et al., 2002).

When the same sound occurs for the second time, it receives the same initial processing than the first sound. However, because the previous sound initiated the formation of a new regularity representation, the relation between this sound and the representation of the first sound is also determined. Their temporal relationship as well as their relationship along the established features is encoded into an episodic representation connecting the two events. Detecting the repetition of a sound could already give rise to the prediction that the next sound will also be the same, as was shown by the elicitation of the MMN response for the 3rd sound of a train that differed from the two previous (identical) sounds (Bendixen, Roeber, & Schröger, 2007). However, in other studies, two repetitions were required before a deviant could trigger a deviance-detection response (Cowan et al., 1993 and Winkler et al., 1996). A possible explanation is that when participants attend to the sounds (as in Bendixen et al.'s study), a single repetition can give rise to a regularity representation. This account is compatible with Bayesian inference rules in model selection (Kersten et al., 2004, Knill and Pouget, 2004 and Yuille and Kersten, 2006) with priors determined by higher-level models, which may be directly related to behavioral goals (and thus would be regarded as voluntary or attentive in the terminology of cognitive psychology).

For unattended sequences, in which MMN was only elicited by a sound after two repetitions of a different sound, the relationship between the 3rd and the 2nd sound is compared with that between the 2nd and the 1st sound. Note that relationships between successive sounds are compared as opposed to representations of individual sounds (Winkler, 2007). When the two relationships are found to be matching, a regularity representation connecting successive sounds is formed. This representation can now predict future events. Enabling the ability of a regularity representation to predict upcoming sounds can also be regarded as an activation process. Taking this notion, we can distinguish "active" and "dormant" regularity representations. Thus an established regularity representation can be dormant (i.e., does not affect the processing of sounds). It can then be activated either by an additional confirming sound event or by attention directed to the sounds (see above). Thus AERS can learn and react fast to emerging patterns in the environment without forsaking prudence (i.e., wasting processing capacity on chance patterns).

Although auditory regularities are detected even when participants focus their attention on a different modality (for reviews, see Haroush et al., 2010 and Sussman, 2007), it is less clear whether strong focusing within the auditory modality can prevent or at least modulate the formation of regularity representations. Modulating effects on deviance detection have been observed by Haroush et al. (2010), whereas Sussman, Winkler, and Wang (2003) showed that the ERP response for deviance in a given feature is suppressed in an unattended sound sequence when participants detect deviants in the same feature in a separate but attended sound stream (for compatible evidence, seeNäätänen et al., 1993, Woldorff et al., 1991 and Woldorff et al., 1998). Either way, attention focused strongly on one stream of sound does not in general prevent the detection of deviations in an unattended stream.

When the model contains at least one (active) regularity representation, predictions for incoming sounds are produced. In our simple case, the model predicts that the next sound is probably identical to the three previous ones. These predictions have an immediate impact on the processing of the input. The sensory memory representations of further sounds are compared with the representation of the sound predicted by the model. In case of a match the regularity representation may be further strengthened. One possible interpretation is that stimuli conforming to the predictions may increase the weight the system attaches to the predictions from the given regularity (i.e., the "confidence" of the system regarding the given prediction). This notion is similar to Friston's predictive coding theory according to which predictions do not only send down contents to the lower level, but also their inferred precision (Feldman & Friston, 2010). Indeed, change-related neural activity has been reported to increase with increasing number of repetitions preceding a change (Javitt, Grochowski, Shelley, & Ritter, 1998). Some studies found a similar effect for the ERP amplitude difference between deviant and standard responses (Bendixen et al., 2007 and Haenschel et al., 2005), although others failed to observe a significant effect (Cowan et al., 1993). On the other hand,Winkler, Karmos, et al.'s (1996) results suggest that stimuli confirming the predictions make the related regularity representations more resistant to elimination. These authors delivered to participants short trains starting with six presentations of tone 'A' followed by 0, 2, 4, or 6 presentations of tone 'B'. The train ended with tone 'C', which differed from both "A" and 'B". Tone 'C' elicited an MMN with respect to tone 'A' even after 4 intervening presentations of tone 'B', showing that the repetition regularity of tone 'A' was not eliminated by repeated presentations of tone 'B'. Although there is no unequivocal proof for the existence of these processes, it stands to reason that regularities, whose predictions are often confirmed, have increased utility for AERS. Haenschel et al. (2005)found an ERP response elicited by regular sounds which increased together with the number of preceding regular sounds (repetition positivity, RP; see also Baldeweg, 2006). The neural process generating RP may be involved in strengthening (sharpening) or making more resistant the corresponding regularity representation(s).

In case a difference is detected between the incoming sound and the prediction from the regularity representation (a prediction error), the stimulus is marked as containing new information. This increases the chance that the stimulus representation receives more detailed processing. The representation of the violated regularity (which predicted the reoccurrence of the same sound in our simple example) is then updated. There is evidence showing that the primary function of the process reflected by the MMN component is related to the regularity representation, as opposed to the deviant sound itself (Winkler & Czigler, 1998). Winkler and Czigler (1998) found that a deviant sound violating two separate regularities within 200 ms elicited two successive MMN responses. In contrast, a deviant sound violating the same regularity two times within 200 ms elicited only a single MMN response. This pattern of results suggests that MMN is primarily related to the regularity violated as opposed to the sound that violates it. We hypothesize that the updating process makes the affected regularity representation (1) carry less weight (confidence) and (2) less resistant to elimination in the future. Several studies (e.g., Winkler et al., 1996 and Winkler et al., 1996) showed that a single deviant does not prevent a consecutive deviant from eliciting MMN. This means that regularity representations are never eliminated by a single non-conforming auditory event. On the other hand, several deviants in a row or a long silent interval following the last regular sound prevent further deviants from eliciting the deviance-related MMN response. The longest silent interval after which a deviant elicited the MMN was found to be ca. 10–12 s (Sams et al., 1993 and Winkler et al., 2001), thus placing an upper bound on the temporal extent of predictions. However, even in these cases, a single regular sound (termed "reminder") can "reactivate" the regularity representation. That is, a deviant sound following a single regular sound (termed the reminder) will again elicit the MMN (for review of the reactivation phenomenon, see Winkler & Cowan, 2005). Reactivation was observed with the reminder separated from the previous regular sound by 30 s (Winkler et al., 2002). Furthermore, reactivation also occurred when the reminder followed six consecutive different deviant sounds (Winkler, Cowan, et al., 1996). Thus it is yet unknown when or how auditory regularity representations are truly eliminated. They only become dormant (not affecting the processing of sounds) by long silent periods or by repeated failure to correctly predict the incoming sound. A similar "dormant" state can be assumed for regularity representations under construction with the third presentation of the standard activating the regularity representation (i.e., making it produce predictions in the future).

Note that once a regularity representation has been established, it starts producing predictions for upcoming sounds in the sequence – that is, it acts as a (possibly partial) generative model of a putative perceptual object (for a description of how some of these "proto-objects" can emerge as perceptual objects that can be consciously experienced, see Section 6.3). However, the build-up of a regularity representation, as described above, is not a predictive process itself. Predictions are about (possible) objects. Therefore, no prediction can be made before a possible object is detected.

It is easy to see, how the system benefits from this mode of operation. Immediate elimination of the regularity representations would be disadvantageous as random fluctuations, which often occur in everyday acoustic environments, and discrete deviant events (exceptions) would reset the system (returning it to the initial non-predicting state), even when the majority of the stimuli follow the detected rule. The existence of a dormant state for regularities further improves the chances of rapidly finding adequate regularity representations for incoming sounds. Together, these features make AERS a robust system in terms of maximizing its predictive capabilities under natural, noisy and variable circumstances.

More Complex Regularities

So far, we focused on how a regularity representation is formed for a sequence of a repeating sound. However, there is evidence that no sound repetition is needed for establishing a regularity representation. For example, if some feature or features are constant within a sequence while other features randomly vary, deviations from the common feature(s) elicit the MMN (Gomes et al., 1995, Huotilainen et al., 1993 and Winkler et al., 1990). From this, we infer that regularity representations have been constructed for the common (invariant) features. Furthermore, when a regularity representation based on feature repetition becomes dormant, it can be reactivated similarly to that described for fully repeating sounds (Ritter, Sussman, Molholm, & Foxe, 2002). This suggests that the relationships between successive sounds are established for each stimulus feature (Nousak et al., 1996 and Ritter et al., 1995).

Identity is a special case of the possible inter-sound relationships. Therefore, one should expect that regularity representations are also established for regular non-repetitive inter-sound relationships. Indeed, this is the case. For example, the regularity of successive sounds continuously increasing or decreasing in pitch is detected similarly to sound repetition (Tervaniemi, Maury, & Näätänen, 1994). Our description of the processes involved in establishing a new regularity representation takes into account this and similar non-repetitive inter-sound relationships. When any detectable relationship between successive sounds repeats (such as, pitch increases from sound 1 to 2 and from sound 2 to 3, etc.) the corresponding regularity representation is activated. Viewed this way, the rising-pitch regularity is no more complex than feature repetition. Studies showing that sounds violating different feature-regularities of the same sound sequence elicit somewhat different MMN responses (e.g., Deacon et al., 1998 and Giard et al., 1995) suggest that several regularities are maintained in parallel in AERS.

In the above discussed regularities, inter-sound relationships repeated immediately (i.e., the length of the repeating cycle was one: the inter-sound relationship was always the same). However, everyday sound sequences often include repeating cycles consisting of several sounds with a characteristic pattern of inter-sound relationships (such as bird trills). Indeed, for example, exchanging two segments within a repetitive cycle of five tones elicits the MMN response (Winkler & Schröger, 1995).

Furthermore, Sussman et al. (Sussman et al., 1998a and Sussman et al., 2002) demonstrated that when a sound sequence is perceived in terms of a repeating pattern, the regularity representations underlying deviance detection are also based on the same pattern. These authors presented repeating cycles consisting of five tones, the first four of which was identical and the last different from them (AAAABAAAAB...). When listeners perceived the repeating cycle, no MMN was elicited by the fifth sound, even though MMN was elicited by this sound when the same sounds were presented in a randomized order or when the listener did not detect the repeating cycle. These results suggest that when the repeating cycle was not present or perceived, the listener's brain treated each sound as a separate unit and predicted the repetition of the more frequent (0.8 probability) sound. However, when the cyclic repetition was detected, the pattern of five sounds became the unit and the "rare" sound was part of this predictable standard.

In order to accommodate repeating cycles with >1 length, we need to extend our previous description of establishing a regularity representation. One possible algorithmic approach assumes that when the formation of a new regularity has been initiated the regularity building process opens a chain of inter-sound relationships (i.e., a sequence of relationships between consecutive sounds). The chain is then either completed when two full repetitions are encountered (establishing a new regularity representation) or discarded when no full repetition is reached before exceeding the capacity of the memory involved in building regularity representations. In support of this hypothesis, Sussman, Ritter, and Vaughan (1998b) found that the five-tone repeating cycle described above is processed in terms of the repeating tone pattern when the presentation rate was sufficiently fast so that two full repetitions of the pattern fit into 10 s. In contrast, regularities of the same sequence were processed in terms of individual sounds, when the presentation rate was slower (cf. Scherg et al., 1989 and Sussman et al., 2002). Further evidence for a temporal capacity limitation in finding repeating cycles has been obtained in studies investigating cyclically repeating noise segments (Kaernbach, 2004).[7] In a recent study, Barascud, Pearce, Griffiths, Friston, and Chait (personal communication by Maria Chait, 10.07.2014) revealed that it is not necessarily needed that the full pattern is repeated in order to for the auditory system to predict its reoccurrence. In one condition studied by Chait and colleagues, the pattern consisted of 10 tones and the listener could detect the regularity after the presentation of only 14 tones as was indicated by an enhancement of the magnetic brain response elicited when the continuation of the pattern was terminated and tones of random pitch were presented instead. This result suggests that the threshold for activating a regularity representation (i.e., making it affect the processing of upcoming sounds) is not two full cycles of the pattern; rather, it is possibly a few (perhaps two or three) hits after the initial pattern has been closed. The simplest form of closure (cf. the gestalt term) comes from encountering again the first sound of the pattern whose representation is being constructed.

There are also regularities which cannot be described by a repeating chain of inter-sound relationships. The simplest example of this type is a sequence with two frequent sounds ('A' and 'B'). Such a sequence includes four frequent inter-sound relationships (A -> A, A -> B, B -> A, and B -> B), each of which is represented. Because, as we argued above, the respective representations (memory traces) are not eliminated by the emergence of other relationship, they can be reinforced by conforming evidence. After a few repetitions of the same relationship, a regularity-representation can be formed. In order to account for this types of regularities, the simplified description of building regularity representations described in the previous paragraphs needs to be extended by relaxing the constrain of regularity building requiring immediate repetition of an inter-sound relationship. Instead, we suggest that repetition must come within the life-time of such traces. Based on studies of testing the temporal limits of rhythm perception (Duke, 1989 and van Norden, 1975), we tentatively suggest that the life-time of these traces is in the order of 1–2 s. Further, there is no specific evidence regarding how many times such an inter-sound relationship must be encountered for the corresponding regularity representation to be activated. Given that the repetitions are not immediate, we assume that more than two recurrence of the same inter-sound relationship is needed.

The regularity representations based on the various frequently encountered inter-sound relationships are built and become active in parallel. Each covers a certain percentage of the incoming sounds, but as long as no inter-sound relationship disappears for long period from the sequence, they will coexist and, as we will show later (see Section 6) by being compatible with each other, they can form a common sound organization. Results showing that even when two sounds have equally high global probability, either one appearing after a longer micro-sequence of the other elicits the MMN (Sams et al., 1983 and Winkler et al., 1992) are compatible with the above description.

The results of many deviance detection studies are compatible with the extended description of forming predictive regularity representations. For example, when pitch increases between the two tones in a tone pair (Ahveninen et al., 2000 and Saarinen et al., 1992) or when tones follow the rule linking different auditory features (Bendixen et al., 2008 and Paavilainen et al., 2007) regularity representations can be formed based on a few frequently occurring inter-sound relationships. Indeed, violating these rules results in the elicitation of MMN.

Finally, recent evidence suggests that regularities can also be extracted for non-adjacent sounds, but only when the sounds intervening between successive sounds of one regularity formed a separate regularity themselves (Bendixen, Schröger, et al., 2012). It should be noted that the paradigm did not allow the two sets of sounds to be segregated by any primitive cue. This is important, because after stream segregation, the sounds forming the two regularities would have become separately adjacent to each other (for other effects related to auditory stream segregation, see Section 6). This result suggests that the auditory system not only registers the relationship between adjacent sounds, but possibly also between non-adjacent ones. However, the fact that such regularities are only detected when the intervening sounds give up a separate regularity suggests that such regularity representations are quite weak normally and require to helping each other to become active. Such help from the other regularity is possible, as these regularities are also compatible with each other (see, again Section 6).

So far, we only considered regularities in which auditory features could be predicted with high accuracy. However, this rarely is the case in real-life situations. Dealing with natural variance involves constructing categories and defining regularities by relationships between stimulus categories instead of between concrete stimuli. Such regularities can be considered as abstract rules as opposed to concrete rules, which are based on relationships between concrete sounds. The categories can be pre-existing (i.e., stored in long-term memory), such as the phoneme set of a language, or episodic, such as for example elements of a bird thrill we get adapted to when spending some time in the vicinity of some birds belonging to the same species. In terms of AERS, categories are distributions of feature values and, as was already described, predictions (a) refer to such distributions and (b) are adapted to the experienced variability of the preceding sound sequence.

Indeed, there is evidence that regularities based on categories are processed within AERS. For example, a generalized version of the pitch-alternation regularity can be constructed if every second tone is set higher in pitch than the preceding one whereas every other tone is set lower than the preceding one. Evidence that this extended regularity of pitch alternation is detected and applied by AERS was obtained by Horváth, Czigler, Sussman, and Winkler (2001; see further evidence for other category-based regularities in Paavilainen et al., 1999, Phillips et al., 2000, Saarinen et al., 1992 and Tervaniemi et al., 2001). There is also evidence that, similarly to concrete regularity representations, representations of abstract regularities can be reactivated (Korzyukov, Winkler, Gumenyuk, & Alho, 2003). It is thus highly likely that concrete and abstract regularity representations are one and the same within AERS. That is, the auditory system is always prepared for constructing representations for "abstract" (non-exact) regularities, treating concrete regularities as abstract ones with very small feature variance, because concrete regularities can seldom be found outside the laboratory.

In summary, it is clearly advantageous for the auditory system to establish representations of the acoustic regularities of the sounds encountered within the environment. Such representations can absorb a large part of the incoming sound, acting as filters for new information (Schröger, 1997, Sinkkonen, 1999 and Winkler et al., 1993). To this end, AERS registers the inter-sound relationships as well as their sequential order. Once the same order of inter-sound relationships has been detected at least a few times (possibly with other relationships intervening), a regularity representation is formed. Thus regularity representations are formed quite fast. In contrast, elimination of these representations is slow. The latter is important, because, it is clear that in any realistic environment, each of the regularity representations of AERS will often fail to correctly predict upcoming sounds, as even the most complex regularity representations cannot capture alone the complexity of a real-life scene. However, as will be discussed in the next sections, continuously summing the predictions of a large set of such simple and fallible regularity representations can produce a robust and flexible representation system, maximizing the predictive power of the model of the acoustic environment.

AUDITORY REGULARITY REPRESENTATIONS, AUDITORY STREAMS, AUDITORY PERCEPTUAL OBJECTS

In every-day auditory environments, most of the time, one encounters several concurrent and intermittent sounds originating from different sources. Due to the physical nature of sounds, the acoustic information arriving from various sources interact with each other within the air according to the mathematical laws applying to waves. If we assume that the auditory system has evolved to find out information about distal objects (the sound sources) and events that gave rise to the sounds, then one of the most important functions of the central auditory system is to break down the incoming signal according to their sources. This is not a trivial task, because there are no simple cues separating the contributions of different sound sources that would work in most situations. In fact, an analysis of natural auditory scenes suggest that there is no unique mathematical solution to finding moving sound sources by the information available to the auditory system (Stoffregen & Bardy, 2001). According to the empiricist point of view (Helmholtz, 1867), the auditory system must, therefore use heuristic computational processes, which are based on assumptions regarding the nature of the sound sources to determine the actual source configuration (see, however, the contrasting view of direct perception; Gibson, 1979). This function has been termed the "auditory scene analysis" by Bregman (1990; for recent reviews, see Ciocca, 2008, Denham and Winkler, 2014, Haykin and Chen, 2005, Shinn-Cunningham and Wang, 2008 and Snyder and Alain, 2007). Many of these assumptions have been described as the laws of perception by the Gestalt school of psychology (Köhler, 1947). Some of them rely on the spectrotemporal configuration of short auditory segments (e.g., co-occurrence of harmonics of a common base and common onset for sounds produced by the same source), but the majority of these constraining assumptions is concerned with the sequential/temporal relationship between sounds coming from a single source (e.g., smooth continuation, common behavior of the sound components originating from the same source, etc.). In order to utilize these principles, the auditory system must store the recent history of the various active sound sources, representing their characteristic acoustic features as well as their dynamic behavior. We propose that the regularity representations described in the previous sections serve also this purpose and that deviance detection and auditory object formation are two tightly interwoven functions of the auditory system.

Before detailing the role of AERS in auditory object formation, here we argue that the features described for the regularity building functions in the previous sections are fully compatible with known properties of auditory streaming, the most widely studied phenomenon within auditory scene analysis (for a detailed analysis, see, Winkler, 2010 and Winkler et al., 2009). The auditory streaming paradigm (van Norden, 1975) consists of a sound sequence mixing together two sets of sounds. The typical sequence takes the form of ABA-ABA-..., where 'A' and 'B' denote two tones differing in frequency and '-' represents a silent period equal to the common duration of the two tones (Fig. 2). Depending on the frequency separation between 'A' and 'B' and the presentation rate (usually characterized by the time between the onsets of consecutive sounds, the stimulus onset asynchrony [SOA]), this sequence is most likely to be experienced either in terms of repeating ABA triplets producing a galloping rhythm in perception (the 'integrated percept') or as two concurrent isochronous streams of sound, a faster paced one consisting of the 'A' and a slower one of the 'B' tones (the 'segregated percept'). Note, however, that other relatively stable percepts are also possible (Denham et al., 2014). With large separation between the two tones and/or fast sound presentation rates, segregation is perceived more commonly, whereas with small frequency separation and/or slow presentation integration is the more common percept (Bregman, 1990 and van Norden, 1975). Streams can be segregated by separation in a variety of auditory features (e.g., Akeroyd et al., 2005, Grimault et al., 2002, Roberts et al., 2002 and Vliegen and Oxenham, 1999) suggesting that auditory streaming is generally based on perceptual dissimilarity (Moore & Gockel, 2002), or rather, taking also into account the effect of presentation rate, auditory streaming is based on rate of perceptual change (Mill et al., 2013 and Winkler et al., 2012). With longer sound sequences, perception inevitably fluctuates between the possible percepts (Anstis and Saida, 1985,Bendixen et al., 2010, Denham et al., 2010, Denham and Winkler, 2006, Leopold and Logothetis, 1999, Pressnitzer and Hupe, 2006, Rahne and Sussman, 2009, Roberts et al., 2002, Schadwinkel and Gutschalk, 2011 and Wessel, 1979). Thus auditory streaming is a multistable perceptual phenomenon (Winkler et al., 2012).

Although perceptual multistability is quite rare under everyday circumstances, this phenomenon is very important for perceptual theories (e.g., Gregory, 1980), as it provides insights into the underlying mechanisms. In short, theories and models of perception need to account for bi-/multistable phenomena (Schwartz, Grimault, Hupe, Moore, & Pressnitzer, 2012). Multistability in auditory streaming suggests that alternative sound organizations are maintained in parallel. Indeed, Horváth et al. (2001) found that in an alternating sequence of two tones, representations for both the regularity of "A is followed by B and vice versa" and also for the rule of "every second tone is A, every other is B" have been maintained in parallel. This was shown by MMN being elicited by violating either one of these rules, only. On this basis, Winkler et al., 2009 and Winkler et al., 2012 suggested that the multistability observed in the auditory streaming paradigm stems from competition between alternative regularity representations describing a sound sequence.

The Auditory Streaming Paradigm

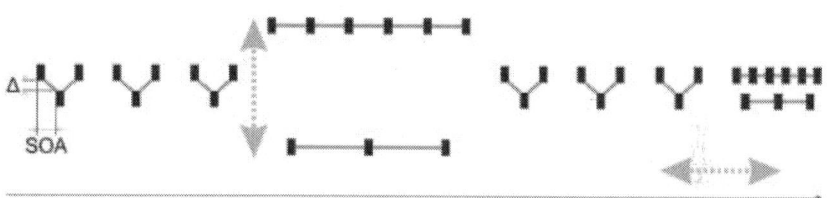

Figure 2. Schematic diagram of the auditory streaming paradigm (van Norden, 1975). Short sounds (depicted by black rectangles) are presented in a repeating ABA-pattern (the horizontal axis marks the passing of time), where A and B denote two sounds differing in at least one stimulus feature, such as the tone frequency (marked by the vertical position). With small feature separation between the two sounds (marked by Δ on the figure) and slow-to-medium presentation rates (marked with the onset-to-onset interval, the Stimulus Onset Asynchrony on the figure), this sequence of sound is typically experienced as a single coherent stream (marked by connecting adjacent sounds with gray lines on the leftmost and the third segment of the figure). However, when feature separation is increased (second segment, the change marked by the dotted gray vertical arrow before the segment) or the presentation rate is increased (rightmost segment, the change marked by the dotted gray horizontal arrow between the third and the rightmost segments), then listeners tend to perceive the sequence as two separate sound streams, each consisting of similar sounds, only (marked by separately connecting the sounds of each stream by gray lines).

On this hypothesis, the properties of auditory regularity representations should be compatible with the phenomena observed for auditory streaming. One can easily draw parallels between the formation of auditory regularity representations (as described in Section 4) and the temporal course of perception at the beginning of an auditory streaming sequence. On the regularity-representation hypothesis, the initial percept is decided by which regularity is discovered first. In the auditory streaming paradigm, with most, but not all combinations of the parameters, the first percept is the integrated one (Denham et al., 2013, Hupe and Pressnitzer, 2012 and Winkler et al., 2012). Once a second regularity is discovered, then competition begins. It has been shown that whereas the choice and duration of the first percept reported for an auditory streaming sequence is highly sensitive to stimulus parameters, the effects of these parameters on the probability and duration of the percepts reported after the first perceptual switch is much more modest (Deike et al., 2012 and Denham et al., 2013). The competition between alternative regularity representations is likely based on adaptation and noise (Mill et al., 2013), as was also suggested for bistable visual phenomena (Shpiro, Moreno-Bote, Rubin, & Rinzel, 2009).

Several studies showed that deviance detection, as indexed by the MMN ERP component, goes hand-in-hand with the segregation of auditory streams. That is, violations of regularities specific to one or another stream only elicit MMN, when separate streams are perceived (e.g., Sussman et al., 1999, Winkler et al., 2006 and Winkler et al., 1993), whereas violations of regularities specific to the whole sequence only elicit the MMN when listeners experience the integrated percept (e.g., Sussman, 2005, Winkler and Cowan, 2005 and Yabe et al., 2001). For example, Winkler et al. (2006) presented tones of intermediate pitch that could join only one of the separate streams formed by the intervening high and low tones, forming different repeating temporal patterns with them. Participants were instructed (and checked on) to voluntarily hold either the high-middle or the low-middle patterns. MMN was elicited when infrequent changes in the timing of the intermediate tones violated the voluntarily held pattern, but not when they violated the possible alternative pattern. Note that deviations only occurred on the tones of intermediate pitch, which were attended all the time. Therefore this effect cannot be explained by attentional filtering. Rather, this is an effect of grouping biased by attention. Furthermore, both bottom-up (Rahne and Sussman, 2009, Winkler et al., 2003 and Winkler et al., 2005) and top-down (Sussman et al., 2002 and Winkler et al., 2006) biasing of the sound organization have parallel effects on deviance detection. The previous example also illustrates a top-down effect on auditory stream segregation. As for a bottom-up effect, Winkler, Sussman, et al. (2003) presented two random tones that intervened between consecutive tones of a repetitive sequence. In one condition, the pitch range of the intervening tones included the pitch of the repeating tone; in the other, the pitch range of the intervening tones was far removed from the pitch of the repeating tone. Occasional intensity deviations of the repeating tone only elicited the MMN in the latter sequences which listeners perceived as segregated into a stream of the repeating tone and a separate stream of the intervening tones. Newborn infants also showed a similar effect, suggesting that this primitive form of stream segregation is already functional at birth (Winkler, Kushnerenko, et al., 2003).

Are auditory streams the true building bricks of sound perception? Cognitive operations are thought to involve objects. Modern theoretical descriptions of auditory objects emphasize similarities between processing principles as opposed to equating features across different modalities (Griffiths and Warren, 2004, Kubovy and van Valkenburg, 2001, Winkler, 2010 and Winkler et al., 2009). Winkler et al. (2009; see, also Winkler, 2010) suggested four defining criteria for (auditory) object representations. Object representations (1) bind together auditory features as well as possibly multiple temporally distinct acoustic events; (2) are separable from other (possibly concurrent) objects; (3) generalize across different instances of the same object; (4) can extrapolate to object parts of which no information reached the senses. The first three criteria are probably self-evident. The last one refers to our experience that even when the information from the distal object that reaches our senses does not cover all parts of the object, the representation formed of this object gives us a reasonable assessment of the missing information (Gregory, 1980). Taking one of Gregory's examples, one almost never sees all four legs of a table. Even so, the table we see does not miss the unseen legs. Because the acoustic signal is ephemeral (i.e., there are no still sounds, which could be revisited at will), the missing information typically awaits us in the future. Therefore, for auditory object representations, the criterion of extrapolation primarily means temporal predictions.

Do auditory streams act as sound objects? Bregman (1990) lists plenty of evidence showing that auditory streams meet the first three of the above-listed criteria for auditory object representations (for a point-by-point listing of psychophysiological evidence, see,Winkler et al., 2009). We suggested that predictive auditory regularity provide the basis for auditory streams. Therefore, we regard auditory streams as auditory object representations.[8]

HOW AERS WORKS WHEN THE MODEL HAS BEEN SET UP

Under everyday circumstances, the model in AERS is almost never "empty"; rather, it contains a mixture of regularity representations that are currently under construction, ones that are active, and ones that are becoming (or already are) dormant, but can still be accessed. The following description of the functioning of AERS discusses how the hypothesized generative models of the auditory environment may be involved in operations necessary for deviance detection as well as in forming auditory streams.

Simultaneous Stream Segregation

Fig. 3 illustrates the sequence of processes establishing auditory sensory representations in AERS up to and including deviance detection. As was discussed in Section 3.1, the first estimation of sound features is marked by the box titled "Initial Feature Analysis". This analysis may already separate components of the input based on frequency and ear of origin, since these information are available from the moment the incoming sounds start affecting the receptor surfaces. A first assessing of the sources is denoted by the box titled "Initial Grouping by Simultaneous Cues". This operation can separate sounds by static features, such as outstanding spectral cues, onset relationship, etc. (Bregman, 1990, de Cheveigné, 2001 and Micheyl and Oxenham, 2010). Establishing segregation by such cues does not require information about previous sounds (represented in the model) but can rather be performed instantaneously using only information from the current sound. For example, two concurrent sounds having distinct narrow frequency bands or different harmonic structures may be separated from each other (Bregman, 1990). An ERP component reflecting segregation by various instantaneous cues has been discovered by Alain and his colleagues (the "Object-Related Negativity", ORN; Alain et al., 2001, Alain et al., 2002,

Hautus and Johnson, 2005, Johnson et al., 2003 and McDonald and Alain, 2005).

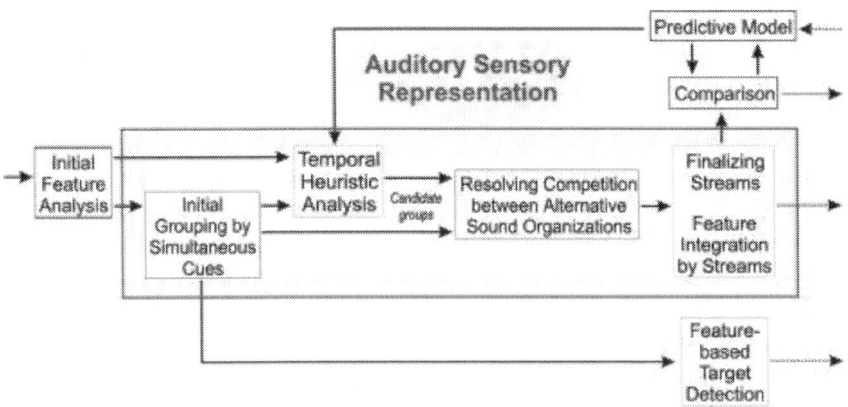

Figure 3. Functional model of the formation of Auditory Sensory Representations. The feature-analyzed sound input enters the processes grouping sounds by (a) Simultaneous Cues and (b) Temporal/Sequential Heuristics. The outcome of initial grouping can serve Target Detection based on a single feature. Also, the same information is passed onto the temporal/sequential grouping process. The two types of grouping processes produce *candidate groups*. Competition between these groups is Resolved, which allows for Finalizing Streams and Integrating Features. The output of Sensory Stimulus Representation is compared with predictions from the Model and Evaluation within the context (see Fig. 1 and Fig. 4).

Grouping/segregation by simultaneous cues occurs within a short time from the onset of a sound and it is one source producing candidates for perceptual sound organization. These candidate groups also provide information for sequential/temporal grouping processes (to be discussed in the next subsection). Finally, these grouping processes allow active monitoring of a given feature and thus the detection of deviance within that feature. This operation is depicted by the box titled "Feature-based Target Detection". As was shown by Näätänen and colleagues, a target feature level can be voluntarily maintained in the brain and target sounds can be detected by comparing all incoming sound to this memory trace (for reviews, see e.g., Näätänen, 1990, Näätänen et al., 2002 and Näätänen et al., 2011). The comparison with a voluntarily maintained memory trace is reflected in an ERP component termed processing negativity (PN). PN is terminated when a sound is found to be different from the target feature or when the target identity is established (reflected by another ERP component, the N2b – seeNäätänen, 1990 and Näätänen and Gaillard, 1983).

Thus separation by instantaneous cues is done early within AERS. However, sounds separable only by temporal/sequential cues or by a combination of two or more features cannot be segregated by these stream-segregation mechanisms.

Temporal Heuristic Analysis

The temporal/sequential heuristic processes hypothesized by Bregman (1990) attempt to account for the contribution of the previously detected active sound sources and identify the onset of new streams from the residual signal. This mode of processing has been termed the old + new strategy. Consequences of the primacy of accounting for the continuation of previously detected streams are demonstrated by the continuity illusion (Riecke et al., 2008) and the phonemic and other auditory restoration effects (Samuel, 1981 and Shinn-Cunningham and Wang, 2008). The heuristic processes summarized in the box titled "Temporal Heuristic Analysis" (Fig. 3) receive their input from initial sound analysis (as suggested by, e.g., the continuity illusion) and from the grouping processes evaluating simultaneous cues. Temporal/sequential grouping processes utilize the predictions generated by the representations of previously detected regularities depicted in the "Predictive Model" component. Temporal/sequential processing occurs probably in parallel with the initial grouping by simultaneous cues as was suggested by the results ofBendixen et al. (2009), who found that during the initial 50 ms following the onset of a fully predictable tone, electrical brain potentials were similar irrespective of whether the sound was actually presented or not. In contrast, the omission of an only temporally predictable tone (i.e., one's whose pitch was not predictable) elicited an ERP response that differed from the response elicited by the actual sound. Compatible results were obtained byFriston and Kiebel's (2009) whose hierarchical dynamic Bayesian model yielded a (simulated) percept for an expected but omitted chirp that mimicked the (simulated) percept when the chirp was actually presented to the model. These results indicated the functioning of an early temporal grouping process utilizing predictions for upcoming sounds.

There is also evidence for interaction between the two types of grouping processes (Bendixen et al., 2010, Dyson and Alain, 2008a, Dyson and Alain, 2008b and Dyson et al., 2005). For example, Bendixen, Jones, et al. (2010) found that the amplitude and scalp topography of the ORN elicited by complex tones with one mistuned partial was modulated by the probability of mistuned complex tones in the sequence. Interactive processing of simultaneous and temporal cues was also demonstrated by studies using behavioral measures (e.g., Ciocca and Darwin, 1999, Darwin et al., 1995, Lee and Shinn-Cunningham, 2008a, Lee and Shinn-Cunningham, 2008b, Steiger and Bregman, 1982 and Teki et al., 2011).

Thus the old + new strategy could be implemented by comparing predictions from the previously detected regularity representations with the incoming sound. Predictions from the model are set on a number of different time scales, in accordance with the diverse temporal basis of the simultaneously active regularity representations. Each of the regularities sets up its own "unit" or temporal chunk of the auditory input. For example, segmental and syllabic units of speech are in the range of 20-80 and 150-300 ms (Poeppel, Idsardi, & van Wassenhove, 2008), whereas the melodic and stress patterns of speech may extend to much longer periods. Thus the analysis of the input must use different temporal chunks or integration periods (Boemio et al., 2005, Grimm and Schröger, 2007, Hickok and Poeppel, 2007, Nelken et al., 2003, Poeppel, 2003 and Poeppel et al., 2008). One well-known integration period, which is ca. 200 ms long, is termed the temporal window of integration (TWI). A large number of perceptual phenomena have been related to this TWI, such as loudness summation, detection masking, etc. (for a review, see Cowan, 1984; for a quantitative model, Zwislocki, 1969). It is thus no surprise that deviance detection also shows effects related to the TWI, such as the detection of omissions from more or less isochronous sound sequences (Yabe et al., 1998) and the integration/separation of closely spaced deviating events (Czigler and Winkler, 1996 and Sussman et al., 1999). Studies contrasting auditory streaming and temporal integration demonstrated that streaming precedes temporal integration (Sussman, 2005 and Yabe et al., 2001). That is, temporal integration occurs within, but not across streams.

Competition and Establishing the Perceived Sound Organization
Bregman (1990) likened the process of selecting one of the alternatives to "voting", where the decomposition of the input that receives the most support from the grouping processes (i.e., many of the grouping processes lead to computing this solution) becomes dominant and it is embraced by the system. This process is marked in Fig. 3 as "Resolving Competition between Alternative Sound Organizations". Considering the structure of this competition, Winkler et al. (2012) suggested that

regularity representations (termed proto-objects by Winkler et al., 2012) compete when they predict the same sound at the same time (termed collision). This local form of competition allow the emergence of full sound organizations (coalitions of proto-objects), which are compatible with each other in the sense Bregman (1990) suggested. Assuming that when two proto-objects collide, they mutually inhibit each other, Mill et al. (2013) showed that compatible proto-objects become "weak/strong" (see below) together in the auditory streaming sequences. That is, whereas the proto-object describing the integrated percept (A-B-A) collides with both of the proto-objects describing the two segregated streams (A---A and B-------B), the latter never collide with each other. Thus, when the integrated proto-object is dominant (i.e., it is perceived), it suppresses both segregated proto-objects; when it is weakened (by adaptation and noise), the two segregated proto-objects become stronger and together they suppress the integrated proto-object (as suggested by Bregman, 1990). One or the other of them becomes dominant (perceived in the foreground) while the other is perceived in the background (i.e., it is not suppressed, as opposed to when the integrated proto-object is dominant). These features of Mill et al. (2013) computational model fully match the perception of the auditory streaming sequences. Further, this notion is also compatible with results suggesting that both redundant and contradictory predictions can be generated for some stimulus sequences (Horváth et al., 2001, Pieszek et al., 2013 and Widmann et al., 2004).

The voting process (competition) is based upon some "strength" (termed activation byMill et al., 2013) measure of the alternative groupings (proto-objects). Strength is provided by the regularity representation supporting the given alternative. It may depend on the type of regularity (representing learning through evolution and individual experience) and also on the "reliability" of each solution: solutions based on regularity representations whose predictions have often been met in the recent past are stronger than alternatives based on regularity representations whose predictions have not always been confirmed by the incoming sounds (for an analysis of the MMN literature on this issue, see Winkler, 2007). Thus prediction error influences the selection between alternative groupings/organizations by decreasing the strength of the related alternatives in the competition. This notion is also compatible with the "model optimization by minimizing prediction errors" principle of predictive coding theories, specifically with a Bayes-optimal (Robert,

2007) variant of model selection (i.e., the winning model provides the greatest evidence or minimum surprise, such as that described by Friston & Kiebel, 2009).[9]

In AERS, the output of the "Finalizing Streams and Feature Integration by Streams" box can be consciously perceived (Fig. 3). In many cases, the solution is (almost) unequivocal (unambiguous auditory scenes). That is, one of the alternatives receives far greater support than any of the others. However, it is also possible that two or more alternative solutions get substantial support from the grouping processes (ambiguous auditory scenes). In this case, perception will fluctuate between the alternatives and, unlike in unambiguous cases, one may voluntarily choose one perception over the other. Thus voting can be biased by top-down effects, but only to a certain degree; that is, one cannot choose an arbitrary solution against an existing dominant stimulus-driven one. This is supported both by results of a large number of behavioral (e.g., van Norden, 1975) and electrophysiological studies (e.g., Sussman et al., 2002 and Winkler et al., 2006).

Finalizing Feature Integration

Once the dominant sound organization is selected, the feature-combinations making up the sounds appearing in the dominant organization are bound together, separately for each of the concurrent sounds, thus creating sound representations, which are inherently linked to auditory streams. Although some influential theories based on visual experiments suggest that feature integration requires focused attention (e.g., Treisman, 1998; see, however, e.g., Duncan and Humphreys, 1989 and Winkler et al., 2005), several studies investigating auditory feature binding found that it can occur even in the absence of focused attention (Gomes et al., 1997, Sussman et al., 1998, Takegata et al., 2001, Takegata et al., 1999, Takegata et al., 2005 and Winkler et al., 2005). However, there is also evidence showing that under some circumstances, the integration of auditory features may not work correctly and illusory feature conjunctions emerge (Hall et al., 2000 and Thompson et al., 2001). For example, when two or more sounds differing both in pitch and timbre are presented in a concurrent array, listeners may identify sounds as being part of the array that have the pitch of one sound and the timbre of another sound from the array. Takegata et al.' (2005) results suggest that correct automatic integration of features occurs also in such cases. Therefore, miscombination of the features may occur during task-related processes. One possible explanation is that listeners use strategies relying on the processes of "feature-based target detection". That is, when the sounds can be segregated by one of the two features, this feature may become the primary cue for the listener in performing a conjunction-search task. A comparison between the results of Thompson et al. (2001) and those of Woods and colleagues (Woods and Alain, 2001 and Woods et al., 1998) supports this interpretation.

At this point in auditory processing, the sensory representation of the incoming sound(s) is complete and it now enters the processes establishing which regularity representations accurately predicted the behavior of the auditory environment ("Comparison" in Fig. 1, Fig. 3 and Fig. 4). Furthermore, the sensory description of the auditory input can now be evaluated with respect to the current goals of the organism (see "Evaluation" in Fig. 1 and Fig. 4) and those parts of the acoustic input for which no prediction existed (residue) can be identified.

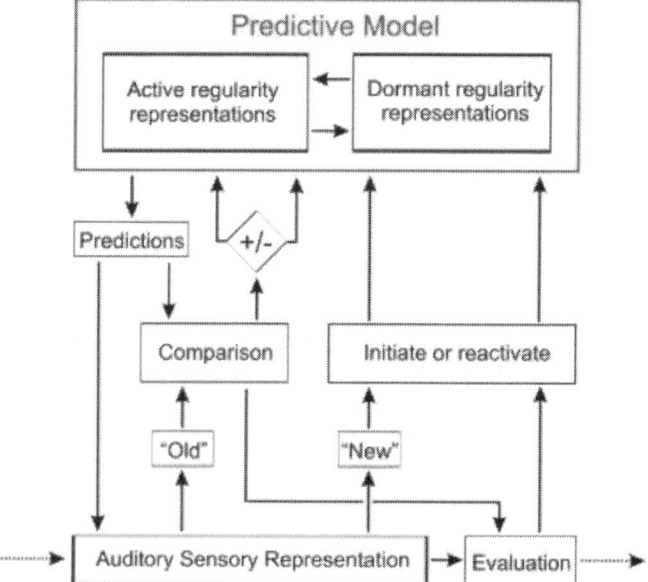

Figure 4. Functional model of sound evaluation and model maintenance. Continuations of streams found during the formation of Auditory Sensory Representations are compared with predictions from the currently Active Regularity Representations of the Model. The outcome of the Comparison process updates the corresponding regularity representations as well as informing the Evaluation process. Sounds for which no predictions existed Initiate the formation of a new regularity representation, which is initially in a dormant state, until its predictive value is confirmed). Dormant regularity representations can also be (re)activated by their behavioral relevance through the evaluation process. In turn, active regularity representation become dormant through updating, if their prediction is violated several times or if no corresponding auditory input is received for a longer period of time (relative to the predicted timing).

Deviance Detection

The output of the feature-integration operation is compared with the predictions produced by the active regularity representations. The outcome of this comparison is used (1) for adjusting the reliability value of the regularity representations (thus affecting their weight in the competition between alternative sound organizations) and (2) for providing the sensory representation of the incoming sound with information about its relationship with the auditory context (i.e., how well it fit the various regularities detected for the acoustic environment). Although the two functions are compatible, as was noted earlier, Winkler and Czigler (1998) showed that the deviance detection signal reflected by MMN is primarily related to adjusting the regularity representations.

In terms of information, deviation from what has been inferred from the generative model represents the new information content of the acoustic event. Thus the sound representations produced by AERS carry with them a description of how much acoustic information they contain. Of crucial importance to the above description of deviance detection is that the sound events separated in the auditory sensory representation are only compared with the regularities that apply to the stream they belong to. This was confirmed by Ritter and colleagues (Ritter et al., 2006 and Ritter et al., 2000), who presented a sound sequence made up from two distinct sets of tones differing from each other on several features, among them tone duration. The two sets of sounds segregated in perception. Occasional sounds with a duration differing from the regular sound durations in either stream only elicited MMN with respect to the stream with which they shared all other auditory features. In contrast, a tone differing in duration from both of two frequently presented tones within the same stream elicits two separate MMNs, one for each of the regular tones (Sussman et al., 2003 and Winkler et al., 1996). Studies delivering sound sequences with some regularity that could only be discovered when the sound sequence was organized in a specific way found that MMN was only elicited, when participants perceived the sound sequence in the given way (Sussman et al., 2007,Sussman et al., 1999, Sussman et al., 2002, Winkler et al., 2003, Winkler et al., 2003 and Winkler et al., 2006). These results strongly support the hypothesis that deviance detection occurs separately within each stream. Furthermore, Bendixen et al., 2010, Bendixen et al., 2013 and Bendixen et al., 2014 found that when a regularity is only discovered when the sounds are organized in a given way, but not within the alternative sound organizations, the presence of this regularity extends the dominance periods of the given organization. Bendixen, Denham, et al. (2010) interleaved two tone sequences separated in pitch. Listeners reported perceiving segregation for longer periods of time when separately, each sequence consisted of repeating tone patterns as compared with when the order of the tones was randomized separately in each sequence. Together, the MMN and perceptual data suggest that repeating patterns are only extracted for streams segregated by other regularities. But once such regularities are discovered, they further support stream segregation and deviance detection.

In summary, the output of AERS not only provides a finely resolved description of the sound event, but also places this event into the context of currently known auditory streams and marks how well it fits the sounds preceding it. That is, the deviance detection system functions as a filter, flagging each sound that carries new information about its source.

Maintenance of the Model

The acoustical environment is in constant flux. New sources become active, whereas active sources may discontinue or change their emission; sources may synchronize or fall out of synchrony. Thus the regular characteristics of the input change all the time. Therefore, the model of the auditory environment requires constant maintenance.

Fig. 4 depicts the functions involved in maintaining and updating the model of the auditory environment. Of crucial importance is the separation of the continuation of streams for which the model already contains representations and the residue, the sounds which did not fit into any of the previously detected streams. When streams are finalized, after competition between the alternative groupings has been resolved, the two types of information are separated. Fig. 4 depicts the two types of information as "Old" (continuation of a known stream) and "New" (residue). The continuation of previously detected streams is used to adjust the regularity representations of the corresponding stream. Just because the given sound has been found to continue the stream, it does not mean that all the regularity representations of the given stream correctly predicted this sound. Therefore predictions for each regularity representation belonging to the stream are compared with the sound continuing the stream in parallel and adjustments are made accordingly. The residue requires the formation of new regularity representations or the reactivation of an existing but dormant one (Fig. 4). This process is modulated by contextual information through the "Evaluation" function. As was already mentioned in Section 4, the representations of those regularities whose predictions are met by the incoming stimulus acquire additional weight as their validity extended in time. In contrast, when a given prediction is not met, then the generative model needs to be updated, as its value for successfully anticipating auditory events has been reduced (Fig. 4). Regularity representations referring to the same sound stream may be fully independent of each other; that is, they describe different aspects of the sound stream. The updating of such regularity representations takes place independently. For example, when two tones with different pitch but uniform duration are alternated, a tone may fit both regularities, violate the constancy of duration, but correctly continue the pitch alternation and vice versa or, violate both regularities. In such cases, multiple regularity violations have been shown to elicit additive MMN components (Alain et al., 1999, Levänen et al., 1993, Schröger, 1995, Schröger, 1996, Takegata et al., 2001,Takegata et al., 2001, Takegata et al., 1999, Winkler and Czigler, 1998 and Winkler et al., 1996). In contrast, violating regularities based on related auditory features, such as multiple temporal or spectral regularities, interfere with each other, typically resulting in subadditivity between the MMN components (Alain et al., 1999, Czigler and Winkler, 1996 and Takegata et al., 2001).

New regularity-representations are created for the newly emerging sound sources, the emission of which shows up as residue after the continuation of the known regularities have been accounted for (Fig. 4). Bendixen et al.'s (2009) results shed some light on the possible timing of the onset of this process. These authors found that when an exact prediction is available for the incoming sound, the ERP response elicited by omitting this sound is not significantly different from that elicited by the sound itself within ca. 50–80 ms from the (expected) onset of the sound. This suggests that the continuation of known regular streams is assessed within this period of time and the residue becomes available by the end of this period. Starting at about this time, a series of auditory cortical ERP components (P1, N1, and P2) can be recorded from the scalp, all of which are sensitive to large acoustic changes in a sequence or sounds presented after long silent periods (Näätänen, 1992 and Näätänen and Picton, 1987). We tentatively suggest that these ERP components may reflect processes involved in building a representation for new auditory objects.[10] In agreement with this notion, a part of the N1 wave is known to be elicited with quite large amplitude by a stimulus delivered after a long silent period then sharply decreasing within the first few sounds of a new train and reaching an asymptotic level after 4–5 sounds (Cowan et al., 1993 and Näätänen and Picton, 1987). This sequence of events is compatible with the assumed phases of establishing a new regularity representation (see Section 4). Furthermore, stimuli eliciting an N1 of high amplitude are likely to capture attention (Näätänen, 1990 and Näätänen et al., 2011), which agrees with one's everyday experience of noticing new sound sources. Also, N1 is increased for familiar sounds (Kirmse, Jacobsen, & Schröger, 2009). Familiar sounds are more likely to enter consciousness (see e.g., hearing one's own name in an unattended channel; Cherry, 1953). Finally, the N1 response is highly sensitive to the direction of focused attention, which is compatible with the assumed top-down influence on detecting new sound sources (Hillyard, Hink, Schwent, & Picton, 1973).

As was described in Section 4, regularity representations can be in a dormant state either because the number of sounds conforming to the given regularity has not yet reached the required level or, because no sounds meeting the regularity have been encountered for some time. The residue may, however, contain a sound conforming to a dormant regularity. In this case, the regularity is reactivated (Winkler & Cowan, 2005; Fig. 4).

Both the formation of new regularity representations and the reactivation of a dormant one can be influenced by top-down processes, including attentional effects (for the interaction of prediction and attention in audition, see Schröger, Marzecová, & SanMiguel, 2015). The relation of the incoming sound to the context is part of the auditory event representation, the outcome of AERS. This includes how well it fit the existing regularities (possible prediction errors) thus allowing evaluation of the sound information with respect to the larger context (including behavioral goals). In response, higher levels of the perceptual/cognitive system can adjust the functioning of AERS by forcing it to look for certain regularities or to reactivate ones, which have become dormant based on stimulus (bottom-up) information alone. Note that the output of the "Comparison" function as specified here is not identical to prediction error in predictive coding models. We separated prediction error into two parts: mismatch between the prediction and the actual continuation of the stream (the "-" branch in the "Old" route; MMN ERP response) and the residue triggering the formation of a new or reactivating a dormant regularity (the "New" route; possibly related to some subcomponent of the N1 ERP response). Within a hierarchical predictive coding model, the former can be addressed locally (within the same level in the hierarchy), whereas the latter requires intervention from higher levels.

Overall, the maintenance of the generative model must ensure that the functional module can quickly adapt to changes in the acoustic environment while keeping its predictive value high all the time. Redundancies in the model, the possibility to reuse outdated regularities, as well as maintaining each of the regularity representations in parallel allows AERS to mark new information for subsequent processing.

COMPARISON WITH EXISTING MODELS OF PREDICTIVE PROCESSING IN PERCEPTION

The assumption of predictive processing is not unique to AERS. For example, predictive modeling theories are based on the same assumption. Garrido and colleagues' (Garrido et al., 2009 and Lieder et al., 2013) and Wacongne and colleagues model (Wacongne, Changeux, & Dehaene, 2012) of ERP responses elicited in the auditory oddball paradigm come closest to the current description. These models have been created to explain the observable MMN response in some specific cases of deviance detection basing on the known neurophysiological properties of the auditory system. However, while these models might be relatively easily extended to cover different auditory features and some additional regularities, as of yet, no attempt has been made to generalize them to the large variety of regularities, whose violation elicits the MMN response. Further, neither model addresses most of the other issues covered by AERS (model build-up, reactivation, separating the updating of existing models from the formation of a new one, or auditory stream segregation, in general) and does not provide a psychological interpretation of the assumed processes. In contrast, AERS covers all of the known regularities, at the cost of relinquishing the neural specificity of the models mentioned above. Thus the two approaches are complementary and may provide synergy in the future. One possible link has been suggested by Winkler and Czigler (2012), who suggested that deviance detection, as reflected by MMN may fit as an intermediate level into a hierarchical predictive coding model. Pre-MMN ERP responses elicited by simpler forms of deviations (for reviews, see Malmierca, Sanchez-Vives, Escera, & Bendixen, 2014, and Grimm & Escera, 2012) may reflect lower levels of the hierarchy.

Kiebel, von Kriegstein, Daunizeau, and Friston (2009) developed a predictive-coding based computational model online recognizing tokens in a hierarchically structured continuous sound. This model has some of the capabilities we assume for the initial build-up of an auditory model including segmentation of continuous sounds, a feature not addressed in (but obviously necessary for) AERS. Thus we regard this model as a possible implementation of some of the functions of AERS. However, again, this model does not consider auditory stream segregation. In contrast, no previous theory or model of auditory stream segregation (Anstis and Saida, 1985, Bregman, 1990, Carlyon, 2004, Jones and Boltz, 1989, Schwartz et al., 2012, Shamma et al., 2011 and Snyder and Alain, 2007) allocates a role for predictive processing. Recently, Mill et al. (2013) based their computational model on the ideas described here; we have already referred to their work in previous sections. Finally, the notion that predictions are an essential aspect of information processing has been considered by modern theorists, such as Bar, 2004, Bar, 2007 and Summerfield and Egner, 2009, and elements of it also appear in Gregory's (1980) and Ahissar and Hochstein's (2004) work. Our description is compatible with many of these ideas. Unfortunately, these general theories are largely based on visual perception, providing little guidance for solving the special problems of auditory perception.

Results of a series of studies suggest that predictive confidence (our term) or precision (Feldman & Friston, 2010) do not fully describe the stimulus-driven determinants of the MMN amplitude. Todd and colleagues (Mullens et al., 2014, Todd et al., 2014, Todd et al., 2014, Todd et al., 2011 and Todd et al., 2013) have repeatedly observed that when the roles of two sounds as frequent standard and rare deviant are periodically exchanged, only the configuration encountered first follows the principle of improved predictive confidence/precision with greater stability of the configuration (i.e., longer periods within which the same standard-deviant configuration remains the same) but not the reversed role configuration, which initially appears after the first role change between the two sounds. The authors refer to this asymmetry as "primacy bias". It probably stems from the different relevance attached to the repetitive and the rare sound by the brain, as the bias can also be manipulated by assigning behavioral relevance to one or the other sound, even when they are first encountered by the listener in a sequence in which the two sounds appear with equal probability (Mullens et al., 2014). If this was the case, then within AERS, primacy bias appears through the evaluation function, which takes into account the contextual relevance attached to each sound by higher-level functions.

AERS ISSUES RELATED TO COMMUNICATION BY SOUND

We started our review by pointing out that communication requires the maintenance of an open channel between the parties. Here we break down this general function for the auditory modality and describe the role of AERS in implementing them in humans.

First, one should consider that often multiple sources are active concurrently and the listener needs to distinguish them in order to being able to follow one (or a few) of these communication channels. Segregating sound sources, including speakers, is the primary function of AERS. As was suggested before, we regard the regularity representations as forming the core of auditory perceptual objects (Winkler et al., 2009). They encode the characteristic auditory features detected for the sound source, allowing the system to determine which part of the incoming sound was likely generated by this source. The output of AERS is the earliest internal sound representation that can be identified, monitored, processed as a speech stream, etc.

Once the sources are distinguished, they can be identified. Identification of sound sources naturally lies outside AERS. However, it affects the AERs as it allows learned information about the given type of source to fine-tune the predictive model. For example, if the source is a bird, we expect chirping sounds (experience with the exact species may allow even more specific predictions). This knowledge can affect AERS via the Evaluation function, biasing the build-up and reactivation of models. Source (speaker) identity is even more crucial for speech perception. Typically, coherent messages come from a single speaker (or, possibly, several speakers speaking in concert – such as a choir; which can be regarded as single sound source). The acoustic features identifying the speaker are encoded in the regularity representations of AERS. This information is relied on by the semantic processes of speech perception. In turn, syntactic and semantic information may allow much sharper predictions for upcoming sounds. Although predictive processes have been hypothesized for language processing, most studies and models consider reading and possibly sign language, but not speech *per se*. However a predictive framework for speech processing has been developed by Kotz and colleagues (Kotz and Schwartze, 2010, Kotz et al., 2009,Rothermich and Kotz, 2013 and Sammler et al., 2010) that is compatible with the idea that AERS regularity representations may be specified by predictions based on syntactic or semantic predictability.

The next issues to be solved are whether the message is directed to us and if so, does it require a response. Although the first information is typically resolved by the general context, at least in infants, the mode of speech (infant vs. adult directed speech) has a significant effect on whether the infant regards him/herself as the addressee of the message (Senju & Csibra, 2008). Prosody generally tells whether the message contains a question. Prosodic regularities are encoded in AERS, as was shown by studies testing prosodic violations (e.g., Honbolygó et al., 2004, Leitman et al., 2009 and Tong et al., 2014). Thus AERS serves as a source of prosodic information to speech processing and, in return, prosodic regularity representations in AERS may be modulated by syntactic information.

Finally, conversations require mutual adaptation from the participants. This includes turn-taking as well as co-adapting their rhythms of speech (Jaffe, Beebe, Feldstein, Crown, & Jasnow, 2001). The temporal aspects of sound sequences, including stimulus rate and rhythmic structures are also represented in AERS. Changes in a regular inter-stimulus interval (e.g., Nordby, Roth, & Pfefferbaum, 1988), train offsets (e.g., Bendixen et al., 2014, Horváth et al., 2010 and Yabe et al., 1998), and violations of higher-order rhythmic (Ladinig, Honing, Háden, & Winkler, 2009) elicit MMN responses. Thus, AERS can play an important role in providing information about response timing.

From our point of view, speech is only one of many possible communication channels. Once the channel is established, the predictions are assumed to foster the processing of the information delivered via the channel. In the case of speech, one can hardly imagine that production and perception can work without generative models; consider, e.g., the German sports reporter Heribert Fassbender, who could articulate up to 26 phonemes per second, and listeners were still being able to comprehend it (N. Blotzki, unpublished Master Thesis, Bonn University). Another benefit of predictions is that they help to detect new acoustic information communicated by the channel: through deviance (irregularity) detection, an inherent property of AERS, we may quickly learn about a change of the speaker or a change in the speaker's state (by detecting changes in voice characteristics).

LIMITATIONS OF AERS AND FUTURE DIRECTIONS

The goal of this review has been to outline a common theoretical framework for conceptualizing phenomena observed in studies of auditory scene analysis and deviance detection. As most theoretical frameworks, the current one is also bound in two different ways: (1) self-imposed limitations regarding the width and depth of the discussion and (2) limitations imposed by the experimental data considered. We already referred to the first in Introduction. We did not review the extensive literature in psychology, acoustics, and neuroscience on early auditory processing. We assume that preprocessed auditory information is available at the input of AERS. Similarly, we did not attempt to outline higher cognitive systems which utilize the information from AERS and can adjust its operation.

It is perhaps more important to consider, how the empirical evidence forming the basis of the current review may limit its scope. Most studies referred in the current review (and also in the literature in general) deal with stimulus configurations, which are far simpler than almost any real-life auditory scene. Specifically, streams in these studies differ from each other by one (in a very few cases two or three) primary auditory features. The same is true for the separation between standard and deviant sounds in most deviance-detection studies. Furthermore, the test sounds are usually short and they seldom fully overlap each other in time; thus promoting discrete sounds as easily discernible units of the incoming sound. Does this mean that the functions and principles described in AERS are only valid for such simplified stimulus configurations? Perhaps not. There is no reason to assume that either auditory streaming or deviance detection would work differently when the sounds are separated by complex spectro-temporal characteristics. For example, Winkler, Teder-Sälejärvi, Horvath, Näätänen, and Sussman (2003) delivered to participants sequences composed of 11 different natural footstep sounds. Ten of these sounds were similar and presented in an approximately even rhythm, thus giving the impression of someone walking. The remaining one sounded as if someone stepped onto a different surface. Although there were no easy-to-define spectral or temporal differences between the two types of footstep sounds, when the different-surface footstep was presented in the 10th position in the footstep sequence, it elicited the MMN. MMN was elicited despite the fact that the participant watched a movie with sounds and street noise was also continuously delivered to the room through loudspeakers. This result demonstrates that the deviance detection system can also utilize complex auditory features within a natural sound environment as for detecting the deviation, the sequence of footsteps had to be segregated from two other continuous streams of sound (street noise and the sound of the movie), both of which covered a wide spectral range, fully overlapping that of the footstep sounds. Thus auditory streams fully overlapping in time and in the spectrum were segregated from each other (for streaming by complex feature differences, see also, Iverson, 1995; for extracting sounds fully embedded in other sounds, see e.g., Chait, 2014, McDermott et al., 2011, Teki et al., 2013 and Teki et al., 2011).

Deviance detection for speech sounds works similarly to other types of sounds with phonetic features showing categorical effects with respect to the languages spoken by the listener (for reviews see, Bishop, 2007, Näätänen, 2001, Pulvermüller and Shtyrov, 2006 and Rimmele et al., 2015). Speaker and speech segregation from noise and from other speakers has been extensively studied in the literature (e.g., Culling and Summerfield, 1995, de Cheveigné et al., 1997 and de Cheveigné et al., 1997; for the engineering point of view see, e.g., Loizou, 2007). However, speech/speaker segregation is typically not an auditory-only function, as understanding speech allows one to sharpen predictions for upcoming sounds. As was already noted in Introduction, such effects are fully compatible with AERS. They are conceptualized as higher-level models in a possible predictive coding hierarchy affecting model selection and parameters in AERS through the "Evaluation" function.

Forming and maintaining regularity representations in AERS do not depend on the feature underlying the regularity. However, AERS does not include mechanisms promoting the emergence of context-specific features (e.g., by plastic changes in the spectro-temporal receptive fields of groups of afferent neurons). It is possible that long-term learning effects shape what features are picked up by our auditory system. Further, although our description focuses on the temporal/sequential cues of auditory stream segregation, we also considered stream segregation by spectral/concurrent cues. As for finding sound units within a realistic auditory scene, together with Nelken et al. (2003), we maintain that sound is analyzed on multiple time scales in parallel, thus allowing parallel formation of regularities based on different units. There exist some computational models capable of segmenting continuous sounds (Coath et al., 2005 and Kiebel et al., 2009). One exciting future direction will be to connect them with computational models based on AERS (Mill et al., 2013). Larger units can then be built from smaller ones, as was reviewed in Section 4.2. Thus the simplified stimulation employed in most studies of auditory stream segregation and deviance detection does not appear to limit the generality of the functions assumed for AERS. On the other hand, we acknowledge that the current description did not explore the effects of task- and knowledge-based strategies on auditory streaming and deviance detection. AERS includes several different ways in which top-down influence can affect its operation. Specifying these effects is an exciting direction for further research (see, e.g., Niessen, van Maanen, & Andringa, 2008).

In summary, AERS is aimed at capturing the intelligent dynamic aspects of auditory perceptual processing, which allows − paraphrasing Köhler's famous sentence − naïve and uncritical listeners to effortlessly experience the auditory world as organized and to select from it meaningful, identifiable objects at will. So, finally, one can say that we listen to sounds through our AERS.

ACKNOWLEDGMENTS

This research was supported by the Hungarian Academy of Sciences (Lendület project, LP2012-36/2012 to IW), by the Reinhart Koselleck grant of the German Research Foundation (Deutsche Forschungsgemeinschaft, DFG, SCH 375/20-1 to ES), and by the Leibniz-Professor Award from the University of Leipzig to IW during the winter term 2003/2004. We thank Johanna Steinberg for proofreading the manuscript.

REFERENCES

1. Ahissar, M., & Hochstein, S. (2004). The reverse hierarchy theory of visual perceptual learning. Trends in Cognitive Sciences, 8(10), 457–464.
2. Ahveninen, J., Jaaskelainen, I. P., Pekkonen, E., Hallberg, A., Hietanen, M., Näätänen, R., et al. (2000). Increased distractibility by task-irrelevant sound changes in abstinent alcoholics. Alcoholism-Clinical and Experimental Research, 24(12), 1850–1854.
3. Akeroyd, M. A., Carlyon, R. P., & Deeks, J. M. (2005). Can dichotic pitches form two streams? Journal of the Acoustical Society of America, 118(2), 977–981.
4. Alain, C. (2007). Breaking the wave: Effects of attention and learning on concurrent sound perception. Hearing Research, 229(1–2), 225–236.
5. Alain, C., Achim, A., & Woods, D. L. (1999). Separate memory-related processing for auditory frequency and patterns. Psychophysiology, 36(6), 737–744.
6. Alain, C., Arnott, S. R., & Picton, T. W. (2001). Bottom-up and top-down influences on auditory scene analysis: Evidence from event-related brain potentials. Journal of Experimental Psychology: Human Perception and Performance, 27(5), 1072–1089.
7. Alain, C., Cortese, F., & Picton, T. W. (1999). Event-related brain activity associated with auditory pattern processing. NeuroReport, 10(11), 2429–2434.
8. Alain, C., Schuler, B. M., & McDonald, K. L. (2002). Neural activity associated with distinguishing concurrent auditory objects. Journal of the Acoustical Society of America, 111(2), 990–995.
9. Andreou, L. V., Kashino, M., & Chait, M. (2011). The role of temporal regularity in auditory segregation. Hearing Research, 280(1–2), 228–235.
10. Anstis, S., & Saida, S. (1985). Adaptation to auditory streaming of frequency modulated tones. Journal of Experimental Psychology: Human Perception and Performance, 11(3), 257–271.
11. Baldeweg, T. (2006). Repetition effects to sounds: Evidence for predictive coding in the auditory system. Trends in Cognitive Sciences, 10(3), 93–94.
12. Baldeweg, T. (2007). ERP repetition effects and mismatch negativity generation – A predictive coding perspective. Journal of Psychophysiology, 21(3–4), 204–213.
13. Bar, M. (2004). Visual objects in context. Nature Reviews Neuroscience, 5(8), 617–629.
14. Bar, M. (2007). The proactive brain: Using analogies and associations to generate predictions. Trends in Cognitive Sciences, 11(7), 280–289.
15. Bendixen, A. (2014). Predictability effects in auditory scene analysis: A review. Frontiers in Neuroscience, 8, 60.
16. Bendixen, A., Bohm, T. M., Szalárdy, O., Mill, R., Denham, S. L., & Winkler, I. (2013). } Different roles of similarity and predictability in auditory stream segregation. Learning and Perception, 5, 37–54.

17. Bendixen, A., Denham, S. L., Gyimesi, K., & Winkler, I. (2010). Regular patterns stabilize auditory streams. Journal of the Acoustical Society of America, 128(6), 3658–3666.
18. Bendixen, A., Denham, S. L., & Winkler, I. (2014). Feature predictability flexibly supports auditory stream segregation or integration. Acta Acustica United with Acustica, 100(5), 888–899.
19. Bendixen, A., Jones, S. J., Klump, G., & Winkler, I. (2010). Probability dependence and functional separation of the object-related and mismatch negativity eventrelated potential components. Neuroimage, 50(1), 285–290.
20. Bendixen, A., Prinz, W. G., Horváth, J., Trujillo-Barreto, N. J., & Schröger, E. (2008). Rapid extraction of auditory feature contingencies. Neuroimage, 41(3), 1111–1119.
21. Bendixen, A., Roeber, U., & Schröger, E. (2007). Regularity extraction and application in dynamic auditory stimulus sequences. Journal of Cognitive Neuroscience, 19(10), 1664–1677.
22. Bendixen, A., SanMiguel, I., & Schröger, E. (2012). Early electrophysiological indicators for predictive processing in audition: A review. International Journal of Psychophysiology, 83(2), 120–131.
23. Bendixen, A., Scharinger, M., Strauss, A., & Obleser, J. (2014). Prediction in the service of comprehension: Modulated early brain responses to omitted speech segments. Cortex, 53, 9–26.
24. Bendixen, A., Schröger, E., Ritter, W., & Winkler, I. (2012). Regularity extraction from non-adjacent sounds. Frontiers in Psychology, 3, 143.
25. Bendixen, A., Schröger, E., & Winkler, I. (2009). I heard that coming: Event-related potential evidence for stimulus-driven prediction in the auditory system. Journal of Neuroscience, 29(26), 8447–8451.
26. Bertrand, O., & Tallon-Baudry, C. (2000). Oscillatory gamma activity in humans: A possible role for object representation. International Journal of Psychophysiology, 38(3), 211–223.
27. Bishop, D. V. M. (2007). Using mismatch negativity to study central auditory processing in developmental language and literacy impairments: Where are we, and where should we be going? Psychological Bulletin, 133(4), 651–672.
28. Boemio, A., Fromm, S., Braun, A., & Poeppel, D. (2005). Hierarchical and asymmetric temporal sensitivity in human auditory cortices. Nature Neuroscience, 8(3), 389–395.
29. Brattico, E., Winkler, I., Näätänen, R., Paavilainen, P., & Tervaniemi, M. (2002). Simultaneous storage of two complex temporal sound patterns in auditory sensory memory. NeuroReport, 13(14), 1747–1751.
30. Bregman, A. S. (1990). Auditory scene analysis. The perceptual organization of sound. Cambridge, MA: MIT Press.
31. Carlyon, R. P. (2004). How the brain separates sounds. Trends in Cognitive Sciences, 8(10), 465–471.

32. Carney, L. H. (2002). Neural basis of audition. In H. Pashler & S. Yantis (Eds.). Stevens' handbook of experimental psychology. Sensation and perception (Vol. 1, pp. 341–396). New York: John Wiley & Sons.
33. Chait, M. (2014). Change detection in complex acoustic scenes. The Journal of the Acoustical Society of America, 135(4), 2171.
34. Chang, E. F. (2014). Feature representation in human speech cortex during perception and production. In E. Budinger (Ed.), Proceedings of the 15th international conference on auditory cortex (pp. 14). Magdeburg: University of Magdeburg.
35. Cherry, E. C. (1953). Some experiments on the recognition of speech, with one and with two ears. Journal of the Acoustical Society of America, 25(5), 975–979.
36. Ciocca, V. (2008). The auditory organization of complex sounds. Frontiers in Bioscience, 13, 148–169.
37. Ciocca, V., & Darwin, C. J. (1999). The integration of nonsimultaneous frequency components into a single virtual pitch. Journal of the Acoustical Society of America, 105(4), 2421–2430.
38. Clark, A. (2013). Whatever next? Predictive brains, situated agents, and the future of cognitive science. Behavioral and Brain Sciences, 36(3), 181–204.
39. Coath, M., Brader, J. M., Fusi, S., & Denham, S. L. (2005). Multiple views of the response of an ensemble of spectro-temporal features support concurrent classification of utterance, prosody, sex and speaker identity. NetworkComputation in Neural Systems, 16(2–3), 285–300.
40. Cowan, N. (1984). On short and long auditory stores. Psychological Bulletin, 96(2), 341–370.
41. Cowan, N. (1987). Auditory sensory storage in relation to the growth of sensation and acoustic information extraction. Journal of Experimental Psychology: Human Perception and Performance, 13(2), 204–215.
42. Cowan, N., Winkler, I., Teder, W., & Näätänen, R. (1993). Memory prerequisites of mismatch negativity in the auditory event-related potential (ERP). Journal of Experimental Psychology: Learning Memory and Cognition, 19(4), 909–921.
43. Culling, J. F., & Summerfield, Q. (1995). Perceptual separation of concurrent speech sounds – Absence of across-frequency grouping by common interaural delay. Journal of the Acoustical Society of America, 98(2), 785–797.
44. Czigler, I., & Winkler, I. (1996). Preattentive auditory change detection relies on unitary sensory memory representations. NeuroReport, 7(15–17), 2413–2417.
45. Darwin, C. J., Hukin, R. W., & Alkhatib, B. Y. (1995). Grouping in pitch perception – Evidence for sequential constraints. Journal of the Acoustical Society of America, 98(2), 880–885.

46. de Cheveigné, A. (2001). The auditory system as a separation machine. In A. J. M. Houtsma, A. Kohlrausch, V. F. Prijs, & R. Schoonhoven (Eds.), Physiological and psychological bases on auditory function (pp. 453–460). Maastricht, The Netherlands: Shaker.

47. de Cheveigné, A., Kawahara, H., Tsuzaki, M., & Aikawa, K. (1997). Concurrent vowel identification. 1. Effects of relative amplitude and F0 difference. Journal of the Acoustical Society of America, 101(5), 2839–2847.

48. de Cheveigné, A., McAdams, S., & Marin, C. M. H. (1997). Concurrent vowel identification. 2. Effects of phase, harmonicity, and task. Journal of the Acoustical Society of America, 101(5), 2848–2856.

49. Deacon, D., Nousak, J. M., Pilotti, M., Ritter, W., & Yang, C. M. (1998). Automatic change detection: Does the auditory system use representations of individual stimulus features or gestalts? Psychophysiology, 35(4), 413–419.

50. Deike, S., Heil, P., Böckmann-Barthel, M., & Brechmann, A. (2012). The build-up of auditory stream segregation: A different perspective. Frontiers in Psychology, 3.

51. Demany, L., & Semal, C. (2008). The role of memory in auditory perception. In W. A. Yost, A. N. Popper, & R. A. Fay (Eds.), Auditory perception of sound sources. Springer handbook of auditory research (pp. 77–113). New York: Springer.

52. Denham, S. L., & Winkler, I. (2014). Auditory perceptual organization. In J. Wagemans (Ed.), Oxford handbook of perceptual organization. http://dx.doi.org/ 10.1093/oxfordhb/9780199686858.013.001 (online publication).

53. Denham, S. L., Bohm, T. M., Bendixen, A., Szalardy, O., Kocsis, Z., Mill, R., et al. (2014). } Stable individual characteristics in the perception of multiple embedded patterns in multistable auditory stimuli. Frontiers in Neuroscience, 8, 25.

54. Denham, S. L., Gyimesi, K., Stefanics, G., & Winkler, I. (2013). Perceptual bistability in auditory streaming: How much do stimulus features matter? Learning & Perception, 5(2), 73–100.

55. Denham, S. L., Gyimesi, K., Stefanics, G., & Winkler, I. (2010). Stability of perceptual organisation in auditory streaming. In E. A. Lopez-Poveda, R. Meddis, & A. R. Palmer (Eds.), The neurophysiological bases of auditory perception (pp. 477–488). New York: Springer.

56. Denham, S. L., & Winkler, I. (2006). The role of predictive models in the formation of auditory streams. Journal of Physiology – Paris, 100(1–3), 154–170.

57. Duke, R. A. (1989). Musicians perception of beat in monotonic stimuli. Journal of Research in Music Education, 37(1), 61–71.

58. Duncan, J., & Humphreys, G. W. (1989). Visual-search and stimulus similarity. Psychological Review, 96(3), 433–458.

59. Dyson, B. J., & Alain, C. (2008a). Is a change as good with a rest? Task-dependent effects of inter-trial contingency on concurrent sound segregation. Brain Research, 1189, 135–144.

60. Dyson, B. J., & Alain, C. (2008b). It all sounds the same to me: Sequential ERP and behavioral effects during pitch and harmonicity judgments. Cognitive Affective & Behavioral Neuroscience, 8(3), 329–343.

61. Dyson, B. J., Alain, C., & He, Y. (2005). Effects of visual attentional load on low-level auditory scene analysis. Cognitive Affective & Behavioral Neuroscience, 5(3), 319–338.

62. Enns, J. T., & Lleras, A. (2008). What's next? New evidence for prediction in human vision. Trends in Cognitive Sciences, 12(9), 327–333.

63. Escera, C., Alho, K., Schröger, E., & Winkler, I. (2000). Involuntary attention and distractibility as evaluated with event-related brain potentials. Audiology and Neuro-Otology, 5(3–4), 151–166.

64. Federmeier, K. D. (2007). Thinking ahead: The role and roots of prediction in language comprehension. Psychophysiology, 44(4), 491–505.

65. Feldman, H., & Friston, K. J. (2010). Attention, uncertainty, and free-energy. Frontiers in Human Neuroscience, 4.

66. Fodor, J. A. (1983). The modularity of mind: An essay on faculty psychology. Cambridge, MA: MIT Press.

67. Formby, D. (1967). Maternal recognition of infants cry. Developmental Medicine and Child Neurology, 9(3), 293. 293-&.

68. Friederici, A. D. (2002). Towards a neural basis of auditory sentence processing. Trends in Cognitive Sciences, 6(2), 78–84.

69. Friedman, D., Cycowicz, Y. M., & Gaeta, H. (2001). The novelty P3: An event-related brain potential (ERP) sign of the brain's evaluation of novelty. Neuroscience and Biobehavioral Reviews, 25(4), 355–373.

70. Friston, K. J. (2005). A theory of cortical responses. Philosophical Transactions of the Royal Society B – Biological Sciences, 360(1456), 815–836.

71. Friston, K. J. (2010). The free-energy principle: A unified brain theory? Nature Reviews Neuroscience, 11(2), 127–138.

72. Friston, K. J., & Kiebel, S. (2009). Cortical circuits for perceptual inference. Neural Networks, 22(8), 1093–1104.

73. Garrido, M. I., Kilner, J. M., Stephan, K. E., & Friston, K. J. (2009). The mismatch negativity: A review of underlying mechanisms. Clinical Neurophysiology, 120(3), 453–463.

74. Ghahramani, Z., & Wolpert, D. M. (1997). Modular decomposition in visuomotor learning. Nature, 386(6623), 392–395.

75. Giard, M. H., Lavikainen, J., Reinikainen, K., Perrin, F., Bertrand, O., Pernier, J., et al. (1995). Separate representation of stimulus frequency, intensity, and duration in auditory sensory memory – An event-related potential and dipole-model analysis. Journal of Cognitive Neuroscience, 7(2), 133–143.

76. Gibson, J. J. (1979). The ecological approach to visual perception. Boston: Houghton Mifflin.

77. Gomes, H., Bernstein, R., Ritter, W., Vaughan, H. G., & Miller, J. (1997). Storage of feature conjunctions in transient auditory memory. Psychophysiology, 34(6), 712–716.

78. Gomes, H., Ritter, W., & Vaughan, H. G. (1995). The nature of preattentive storage in the auditory-system. Journal of Cognitive Neuroscience, 7(1), 81–94.

79. Gregory, R. L. (1980). Perceptions as hypotheses. Philosophical Transactions of the Royal Society of London, Series B: Biological Sciences, 290(1038), 181–197.

80. Griffiths, T. D., & Warren, J. D. (2004). What is an auditory object? Nature Reviews Neuroscience, 5(11), 887–892.

81. Grimault, N., Bacon, S. P., & Micheyl, C. (2002). Auditory stream segregation on the basis of amplitude-modulation rate. Journal of the Acoustical Society of America, 111(3), 1340–1348.

82. Grimm, S., & Escera, C. (2012). Auditory deviance detection revisited: Evidence for a hierarchical novelty system. International Journal of Psychophysiology, 85(1), 88–92.

83. Grimm, S., & Schröger, E. (2007). The processing of frequency deviations within sounds: Evidence for the predictive nature of the mismatch negativity (MMN) system. Restorative Neurology and Neuroscience, 25(3–4), 241–249.

84. Haenschel, C., Vernon, D. J., Dwivedi, P., Gruzelier, J. H., & Baldeweg, T. (2005). Event-related brain potential correlates of human auditory sensory memorytrace formation. Journal of Neuroscience, 25(45), 10494–10501.

85. Hagoort, P. (2008). The fractionation of spoken language understanding by measuring electrical and magnetic brain signals. Philosophical Transactions of the Royal Society B – Biological Sciences, 363(1493), 1055–1069.

86. Hall, M. D., Pastore, R. E., Acker, B. E., & Huang, W. Y. (2000). Evidence for auditory feature integration with spatially distributed items. Perception & Psychophysics, 62(6), 1243–1257.

87. Haroush, K., Hochstein, S., & Deouell, L. Y. (2010). Momentary fluctuations in allocation of attention: Cross-modal effects of visual task load on auditory discrimination. Journal of Cognitive Neuroscience, 22(7), 1440–1451.

88. Hautus, M. J., & Johnson, B. W. (2005). Object-related brain potentials associated with the perceptual segregation of a dichotically embedded pitch. Journal of the Acoustical Society of America, 117(1), 275–280.

89. Haykin, S., & Chen, Z. (2005). The cocktail party problem. Neural Computation, 17(9), 1875–1902.

90. Helmholtz, H. v. (1867). Handbuch der physiologischen Optik. Leipzig: Voss.

91. Hickok, G., & Poeppel, D. (2007). Opinion – The cortical organization of speech processing. Nature Reviews Neuroscience, 8(5), 393–402.

92. Hillyard, S. A., Hink, R. F., Schwent, V. L., & Picton, T. W. (1973). Electrical signs of selective attention in human brain. Science, 182(4108), 177–180.

93. Hohwy, J. (2007). Functional integration and the mind. Synthese, 159(3), 315–328.

94. Hommel, B., Musseler, J., Aschersleben, G., & Prinz, W. (2001). The theory of event coding (TEC): A framework for perception and action planning. Behavioral and Brain Sciences, 24(5), 849–878.

95. Honbolygó, F., Csépe, V., & Ragó, A. (2004). Suprasegmental speech cues are automatically processed by the human brain: A mismatch negativity study. Neuroscience Letters, 363(1), 84–88.

96. Horváth, J., Czigler, I., Jacobsen, T., Maeß, B., Schröger, E., & Winkler, I. (2008). MMN or no MMN: No magnitude of deviance effect on the MMN amplitude. Psychophysiology, 45(1), 60–69.

97. Horváth, J., Czigler, I., Sussman, E., & Winkler, I. (2001). Simultaneously active preattentive representations of local and global rules for sound sequences in the human brain. Cognitive Brain Research, 12(1), 131–144.

98. Horváth, J., Müller, D., Weise, A., & Schröger, E. (2010). Omission mismatch negativity builds up late. NeuroReport, 21(7), 537–541.

99. Horváth, J., Winkler, I., & Bendixen, A. (2008). Do N1/MMN, P3a, and RON form a strongly coupled chain reflecting the three stages of auditory distraction? Biological Psychology, 79(2), 139–147.

100. Hosemann, J., Herrmann, A., Steinbach, M., Bornkessel-Schlesewsky, I., & Schlesewsky, M. (2013). Lexical prediction via forward models: N400 evidence from German Sign Language. Neuropsychologia, 51(11), 2224–2237.

101. Huotilainen, M., Ilmoniemi, R. J., Lavikainen, J., Tiitinen, H., Alho, K., Sinkkonen, J., et al. (1993). Interaction between representations of different features of auditory sensory memory. NeuroReport, 4(11), 1279–1281.

102. Hupe, J. M., & Pressnitzer, D. (2012). The initial phase of auditory and visual scene analysis. Philosophical Transactions of the Royal Society B – Biological Sciences, 367(1591), 942–953.

103. Iverson, P. (1995). Auditory stream segregation by musical timbre: Effects of static and dynamic acoustic attributes. Journal of Experimental Psychology: Human Perception and Performance, 21(4), 751–763.

104. Jacobsen, T., Horvath, J., Schröger, E., Lattner, S., Widmann, A., & Winkler, I. (2004). Pre-attentive auditory processing of lexicality. Brain and Language, 88(1), 54–67.

105. Jaffe, J., Beebe, B., Feldstein, S., Crown, C. L., & Jasnow, M. D. (2001). Rhythms of dialogue in infancy: Coordinated timing in development. Monographs of the Society for Research in Child Development, 66(2), i–viii, 1–132.

106. Javitt, D. C., Grochowski, S., Shelley, A. M., & Ritter, W. (1998). Impaired mismatch negativity (MMN) generation in schizophrenia as a function of stimulus deviance, probability, and interstimulus/interdeviant interval. Evoked Potentials-Electroencephalography and Clinical Neurophysiology, 108(2), 143–153.

107. Johnson, B. W., Hautus, M., & Clapp, W. C. (2003). Neural activity associated with binaural processes for the perceptual segregation of pitch. Clinical Neurophysiology, 114(12), 2245–2250.

108. Jones, M. R. (1976). Time, our lost dimension: Toward a new theory of perception, attention, and memory. Psychological Review, 83(5), 323–355.

109. Jones, M. R., & Boltz, M. (1989). Dynamic attending and responses to time. Psychological Review, 96(3), 459–491.

110. Kaernbach, C. (2004). The memory of noise. Experimental Psychology, 51(4), 240–248.

111. Kersten, D., Mamassian, P., & Yuille, A. (2004). Object perception as Bayesian inference. Annual Review of Psychology, 55, 271–304.

112. Kiebel, S. J., von Kriegstein, K., Daunizeau, J., & Friston, K. J. (2009). Recognizing sequences of sequences. PLoS Computational Biology, 5(8), e1000464.

113. Kirmse, U., Jacobsen, T., & Schröger, E. (2009). Familiarity affects environmental sound processing outside the focus of attention: An event-related potential study. Clinical Neurophysiology, 120(5), 887–896.

114. Knill, D. C., & Pouget, A. (2004). The Bayesian brain: The role of uncertainty in neural coding and computation. Trends in Neurosciences, 27(12), 712–719.

115. Köhler, W. (1947). Gestalt psychology: An introduction to new concepts in modern psychology. New York: Liveright Publishing.

116. Korzyukov, O. A., Winkler, I., Gumenyuk, V. I., & Alho, K. (2003). Processing abstract auditory features in the human auditory cortex. Neuroimage, 20(4), 2245–2258.

117. Kotz, S. A., & Schwartze, M. (2010). Cortical speech processing unplugged: A timely subcortico-cortical framework. Trends in Cognitive Sciences, 14(9), 392–399.

118. Kotz, S. A., Schwartze, M., & Schmidt-Kassow, M. (2009). Non-motor basal ganglia functions: A review and proposal for a model of sensory predictability in auditory language perception. Cortex, 45(8), 982–990.

119. Kubovy, M., & van Valkenburg, D. (2001). Auditory and visual objects. Cognition, 80(1–2), 97–126.

120. Kujala, T., Tervaniemi, M., & Schröger, E. (2007). The mismatch negativity in cognitive and clinical neuroscience: Theoretical and methodological considerations. Biological Psychology, 74(1), 1–19.

121. Ladinig, O., Honing, H., Háden, G., & Winkler, I. (2009). Probing attentive and preattentive emergent meter in adult listeners without extensive music training. Music Perception, 26(4), 377–386.

122. Lee, A. K. C., & Shinn-Cunningham, B. G. (2008a). Effects of frequency disparities on trading of an ambiguous tone between two competing auditory objects. Journal of the Acoustical Society of America, 123(6), 4340–4351.

123. Lee, A. K. C., & Shinn-Cunningham, B. G. (2008b). Effects of reverberant spatial cues on attention-dependent object formation. JARO – Journal of the Association for Research in Otolaryngology, 9(1), 150–160.

124. Leitman, D. I., Sehatpour, P., Shpaner, M., Foxe, J. J., & Javitt, D. C. (2009). Mismatch negativity to tonal contours suggests preattentive perception of prosodic content. Brain Imaging and Behavior, 3(3), 284–291.

125. Leonard, M. K., & Chang, E. F. (2014). Dynamic speech representations in the human temporal lobe. Trends in Cognitive Sciences, 18(9), 472–479.

126. Leopold, D. A., & Logothetis, N. K. (1999). Multistable phenomena: Changing views in perception. Trends in Cognitive Sciences, 3(7), 254–264.

127. Levänen, S., Hari, R., McEvoy, L., & Sams, M. (1993). Responses of the human auditory-cortex to changes in one versus two stimulus features. Experimental

128. Brain Research, 97(1), 177–183.

129. Lieder, F., Stephan, K. E., Daunizeau, J., Garrido, M. I., & Friston, K. J. (2013). A neurocomputational model of the mismatch negativity. PLoS Computational Biology, 9(11).

130. Loizou, P. C. (2007). Speech enhancement: Theory and practice. Boca Raton, FL: CRC Press.

131. Malmierca, M. S., Sanchez-Vives, M. V., Escera, C., & Bendixen, A. (2014). Neuronal adaptation, novelty detection and regularity encoding in audition. Frontiers in Systems Neuroscience, 8.

132. McDermott, J. H., Wrobleski, D., & Oxenham, A. J. (2011). Recovering sound sources from embedded repetition. Proceedings of the National academy of Sciences of the United States of America, 108(3), 1188–1193.

133. McDonald, K. L., & Alain, C. (2005). Contribution of harmonicity and location to auditory object formation in free field: Evidence from event-related brain potentials. Journal of the Acoustical Society of America, 118(3), 1593–1604.

134. Micheyl, C., & Oxenham, A. J. (2010). Pitch, harmonicity and concurrent sound segregation: Psychoacoustical and neurophysiological findings. Hearing Research, 266(1–2), 36–51.

135. Mill, R. W., Bohm, T. M., Bendixen, A., Winkler, I., & Denham, S. L. (2013). Modelling } the emergence and dynamics of perceptual organisation in auditory streaming. PLoS Computational Biology, 9(3).

136. Moore, B. C. J., & Gockel, H. (2002). Factors influencing sequential stream segregation. Acta Acustica United with Acustica, 88(3), 320–333.

137. Mullens, D., Woodley, J., Whitson, L., Provost, A., Heathcote, A., Winkler, I., et al. (2014). Altering the primacy bias – How does a prior task affect mismatch negativity? Psychophysiology, 51(5), 437–445.

138. Mumford, D. (1992). On the computational architecture of the neocortex II. The role of cortico-cortical loops. Biological Cybernetics, 66(3), 241–251.

139. Näätänen, R. (1990). The role of attention in auditory information-processing as revealed by event-related potentials and other brain measures of cognitive function. Behavioral and Brain Sciences, 13(2), 201–232.

140. Näätänen, R. (1992). Attention and brain function. Hillsdale, NJ: Erlbaum.
141. Näätänen, R. (2001). The perception of speech sounds by the human brain as reflected by the mismatch negativity (MMN) and its magnetic equivalent (MMNm). Psychophysiology, 38(1), 1–21.
142. Näätänen, R., & Alho, K. (1997). Mismatch negativity – The measure for central sound representation accuracy. Audiology and Neuro-Otology, 2(5), 341–353.
143. Näätänen, R., Alho, K., & Schröger, E. (2002). Electrophysiology of attention. In H. Pashler & J. Wixted (Eds.). Stevens' handbook of experimental psychology. Methodology in experimental psychology (Vol. 4, 3rd ed., pp. 601–653). New York: John Wiley.
144. Näätänen, R., & Gaillard, A. W. K. (1983). The orienting reflex and the N2 deflection of the event-related potential (ERP). In A. W. K. Gaillard & W. Ritter (Eds.), Tutorials in event related potential research: Endogenous components (pp. 119–141). Amsterdam: Elsevier Science Ltd.
145. Näätänen, R., Gaillard, A. W. K., & Mäntysalo, S. (1978). Early selective-attention effect on evoked potential reinterpreted. Acta Psychologica, 42, 313–329.
146. Näätänen, R., Kujala, T., & Winkler, I. (2011). Auditory processing that leads to conscious perception: A unique window to central auditory processing opened by the mismatch negativity and related responses. Psychophysiology, 48(1), 4–22.
147. Näätänen, R., Lehtokoski, A., Lennes, M., Cheour, M., Huotilainen, M., Iivonen, A., et al. (1997). Language-specific phoneme representations revealed by electric and magnetic brain responses. Nature, 385(6615), 432–434.
148. Näätänen, R., Paavilainen, P., Tiitinen, H., Jiang, D., & Alho, K. (1993). Attention and mismatch negativity. Psychophysiology, 30(5), 436–450.
149. Näätänen, R., & Picton, T. (1987). The N1 wave of the human electric and magnetic response to sound – A review and an analysis of the component structure. Psychophysiology, 24(4), 375–425.
150. Näätänen, R. (1984). In search of a short duration memory trace of a stimulus in the human brain. In L. Pulkkinen & P. Lyytinen (Eds.), Human action and personality. Essays in honor of Martti Takala. Jyväskylä studies in education, psychology and social research 54 (pp. 29–43). Jyväskylä: University of Jyväskylä.
151. Näätänen, R., Tervaniemi, M., Sussman, E., Paavilainen, P., & Winkler, I. (2001). 'Primitive intelligence' in the auditory cortex. Trends in Neurosciences, 24(5), 283–288.
152. Näätänen, R., & Winkler, I. (1999). The concept of auditory stimulus representation in cognitive neuroscience. Psychological Bulletin, 125(6), 826–859.

153. Nelken, I., Fishbach, A., Las, L., Ulanovsky, N., & Farkas, D. (2003). Primary auditory cortex of cats: Feature detection or something else? Biological Cybernetics, 89(5), 397–406.

154. Niessen, M. E., van Maanen, L., & Andringa, T. C. (2008). Disambiguating sound through context. International Journal of Semantic Computing, 2(3), 327–341.

155. Nordby, H., Roth, W. T., & Pfefferbaum, A. (1988). Event-related potentials to timedeviant and pitch-deviant tones. Psychophysiology, 25(3), 249–261.

156. Nousak, J. M. K., Deacon, D., Ritter, W., & Vaughan, H. G. (1996). Storage of information in transient auditory memory. Cognitive Brain Research, 4(4), 305–317.

157. Opitz, B., Schröger, E., & von Cramon, D. Y. (2005). Sensory and cognitive mechanisms for preattentive change detection in auditory cortex. European Journal of Neuroscience, 21(2), 531–535.

158. Paavilainen, P., Arajärvi, P., & Takegata, R. (2007). Preattentive detection of
159. nonsalient contingencies between auditory features. NeuroReport, 18(2), 159–163.

160. Paavilainen, P., Jaramillo, M., Näätänen, R., & Winkler, I. (1999). Neuronal populations in the human brain extracting invariant relationships from acoustic variance. Neuroscience Letters, 265(3), 179–182.

161. Perrin, F., Garcia-Larrea, L., Mauguiere, F., & Bastuji, H. (1999). A differential brain response to the subject's own name persists during sleep. Clinical Neurophysiology, 110(12), 2153–2164.

162. Phillips, C., Pellathy, T., Marantz, A., Yellin, E., Wexler, K., Poeppel, D., et al. (2000). Auditory cortex accesses phonological categories: An MEG mismatch study. Journal of Cognitive Neuroscience, 12(6), 1038–1055.

163. Pieszek, M., Widmann, A., Gruber, T., & Schröger, E. (2013). The human brain maintains contradictory and redundant auditory sensory predictions. PLoS One, 8(1).

164. Poeppel, D. (2003). The analysis of speech in different temporal integration windows: Cerebral lateralization as 'asymmetric sampling in time'. Speech Communication, 41(1), 245–255.

165. Poeppel, D., Idsardi, W. J., & van Wassenhove, V. (2008). Speech perception at the interface of neurobiology and linguistics. Philosophical Transactions of the Royal Society B – Biological Sciences, 363(1493), 1071–1086.

166. Polich, J. (2007). Updating p300: An integrative theory of P3a and P3b. Clinical Neurophysiology, 118(10), 2128–2148.

167. Pressnitzer, D., & Hupe, J. M. (2006). Temporal dynamics of auditory and visual bistability reveal common principles of perceptual organization. Current Biology, 16(13), 1351–1357.

168. Pulvermüller, F., Kujala, T., Shtyrov, Y., Simola, J., Tiitinen, H., Alku, P., et al. (2001). Memory traces for words as revealed by the mismatch negativity. Neuroimage, 14(3), 607–616.

169. Pulvermüller, F., & Shtyrov, Y. (2006). Language outside the focus of attention: The mismatch negativity as a tool for studying higher cognitive processes. Progress in Neurobiology, 79(1), 49–71.

170. Rahne, T., & Sussman, E. (2009). Neural representations of auditory input accommodate to the context in a dynamically changing acoustic environment. European Journal of Neuroscience, 29(1), 205–211.

171. Rao, R. P. N., & Ballard, D. H. (1999). Predictive coding in the visual cortex: A functional interpretation of some extra-classical receptive-field effects. Nature Neuroscience, 2(1), 79–87.

172. Riecke, L., van Opstal, A. J., & Formisano, E. (2008). The auditory continuity illusion: A parametric investigation and filter model. Perception & Psychophysics, 70(1), 1–12.

173. Rimmele, J., Schröger, E., & Bendixen, A. (2012). Age-related changes in the use of regular patterns for auditory scene analysis. Hearing Research, 289(1–2), 98–107.

174. Rimmele, J., Sussman, E., & Poeppel, D. (2015). The role of temporal structure in the investigation of sensory memory, auditory scene analysis, and speech perception: A healthy-aging perspective. International Journal of Psychophysiology, 95(2), 175–183.

175. Rinne, T., Sarkka, A., Degerman, A., Schröger, E., & Alho, K. (2006). Two separate mechanisms underlie auditory change detection and involuntary control of attention. Brain Research, 1077, 135–143.

176. Ritter, W., de Sanctis, P., Molholm, S., Javitt, D. C., & Foxe, J. J. (2006). Preattentively grouped tones do not elicit MMN with respect to each other. Psychophysiology, 43(5), 423–430.

177. Ritter, W., Deacon, D., Gomes, H., Javitt, D. C., & Vaughan, H. G. (1995). The mismatch negativity of event-related potentials as a probe of transient auditory memory – A review. Ear and Hearing, 16(1), 52–67.

178. Ritter, W., Sussman, E., & Molholm, S. (2000). Evidence that the mismatch negativity system works on the basis of objects. NeuroReport, 11(1), 61–63.

179. Ritter, W., Sussman, E., Molholm, S., & Foxe, J. J. (2002). Memory reactivation or reinstatement and the mismatch negativity. Psychophysiology, 39(2), 158–165.

180. Robert, C. P. (2007). The Bayesian choice: From decision-theoretic foundations to computational implementation. New York: Springer.

181. Roberts, B., Glasberg, B. R., & Moore, B. C. J. (2002). Primitive stream segregation of tone sequences without differences in fundamental frequency or passband. Journal of the Acoustical Society of America, 112(5), 2074–2085.

182. Rothermich, K., & Kotz, S. A. (2013). Predictions in speech comprehension: fMRI evidence on the meter-semantic interface. Neuroimage, 70, 89–100.

183. Roye, A., Jacobsen, T., & Schröger, E. (2007). Personal significance is encoded automatically by the human brain: An event-related potential study with ringtones. European Journal of Neuroscience, 26(3), 784–790.

184. Saarinen, J., Paavilainen, P., Schröger, E., Tervaniemi, M., & Näätänen, R. (1992). Representation of abstract attributes of auditory-stimuli in the human brain. NeuroReport, 3(12), 1149–1151.
185. Sammler, D., Kotz, S. A., Eckstein, K., Ott, D. V. M., & Friederici, A. D. (2010). Prosody meets syntax: The role of the corpus callosum. Brain, 133, 2643–2655.
186. Sams, M., Alho, K., & Näätänen, R. (1983). Sequential effects in the ERP in discriminating two stimuli. Biological Psychology, 17(1), 41–58.
187. Sams, M., Hari, R., Rif, J., & Knuutila, J. (1993). The human auditory sensory memory trace persists about 10 sec – Neuromagnetic evidence. Journal of Cognitive Neuroscience, 5(3), 363–370.
188. Samuel, A. G. (1981). The role of bottom-up confirmation in the phonemic restoration illusion. Journal of Experimental Psychology: Human Perception and Performance, 7(5), 1124–1131.
189. Schadwinkel, S., & Gutschalk, A. (2011). Transient bold activity locked to perceptual reversals of auditory streaming in human auditory cortex and inferior colliculus. Journal of Neurophysiology, 105(5), 1977–1983.
190. Scherg, M., Vajsar, J., & Picton, T. W. (1989). A source analysis of the late human auditory evoked potentials. Journal of Cognitive Neuroscience, 1, 336–355.
191. Schofield, B. R. (2010). Structural organization of the descending auditory pathway. In A. IRees & A. R. Palmer (Eds.), The Oxford handbook of auditory science: The auditory brain (pp. 43–64). Oxford: Oxford University Press.
192. Schönwiesner, M., & Zatorre, R. J. (2009). Spectro-temporal modulation transfer function of single voxels in the human auditory cortex measured with highresolution fMRI. Proceedings of the National academy of Sciences of the United States of America, 106(34), 14611–14616.
193. Schröger, E. (1995). Processing of auditory deviants with changes in one-stimulus versus two-stimulus dimensions. Psychophysiology, 32(1), 55–65.
194. Schröger, E. (1996). Interaural time and level differences: Integrated or separated processing? Hearing Research, 96(1–2), 191–198.
195. Schröger, E. (1997). On the detection of auditory deviations: A pre-attentive activation model. Psychophysiology, 34(3), 245–257.
196. Schröger, E. (2007). Mismatch negativity – A microphone into auditory memory. Journal of Psychophysiology, 21(3–4), 138–146.
197. Schröger, E., Bendixen, A., Denham, S. L., Mill, R. W., Böhm, T. M., & Winkler, I. (2014). Predictive regularity representations in violation detection and auditory stream segregation: From conceptual to computational models. Brain Topography, 27, 565–577.
198. Schröger, E., Marzecová, A., & SanMiguel, I. (2015). Attention and prediction in human audition: A lesson from cognitive psychophysiology. European Journal of Neuroscience, 41(5), 641–664.

199. Schröger, E., Tervaniemi, M., & Huotilainen, M. (2004). Bottom-up and top-down flows of information within auditory memory: Electrophysiological evidence. In C. Kaernbach, E. Schröger, & H. J. Müller (Eds.), Psychophysics beyond sensation: Laws and invariants of human cognition. Scientific psychology series (pp. 389–407). Mahwah, NJ: Lawrence Erlbaum Associates.

200. Schröger, E., & Winkler, I. (1995). Presentation rate and magnitude of stimulus deviance effects on human pre-attentive change detection. Neuroscience Letters, 193(3), 185–188.

201. Schubotz, R. I. (2007). Prediction of external events with our motor system: Towards a new framework. Trends in Cognitive Sciences, 11(5), 211–218.

202. Schwartz, J. L., Grimault, N., Hupe, J. M., Moore, B. C. J., & Pressnitzer, D. (2012). Multistability in perception: Binding sensory modalities, an overview. Philosophical Transactions of the Royal Society B – Biological Sciences, 367(1591), 896–905.

203. Schwartze, M., Tavano, A., Schröger, E., & Kotz, S. A. (2012). Temporal aspects of prediction in audition: Cortical and subcortical neural mechanisms. International Journal of Psychophysiology, 83(2), 200–207.

204. Senju, A., & Csibra, G. (2008). Gaze following in human infants depends on communicative signals. Current Biology, 18(9), 668–671.

205. Shamma, S. A., Elhilali, M., & Micheyl, C. (2011). Temporal coherence and attention in auditory scene analysis. Trends in Neurosciences, 34(3), 114–123.

206. Shinn-Cunningham, B. G., & Wang, D. (2008). Influences of auditory object formation on phonemic restoration. Journal of the Acoustical Society of America, 123(1), 295–301.

207. Shpiro, A., Moreno-Bote, R., Rubin, N., & Rinzel, J. (2009). Balance between noise and adaptation in competition models of perceptual bistability. Journal of Computational Neuroscience, 27(1), 37–54.

208. Siddle, D. A. T. (1991). Orienting, habituation, and resource-allocation – An associative analysis. Psychophysiology, 28(3), 245–259.

209. Sinkkonen, J. (1999). Information and resource allocation. In R. Baddeley, P. Hancock, & P. Foldiak (Eds.), Information theory and the brain (pp. 241–254). Cambridge: Cambridge University Press.

210. Snyder, J. S., & Alain, C. (2007). Toward a neurophysiological theory of auditory stream segregation. Psychological Bulletin, 133(5), 780–799.

211. Sokolov, E. N. (1963). Higher nervous functions: The orienting reflex. Annual Review of Physiology, 25, 545–580.

212. Steiger, H., & Bregman, A. S. (1982). Competition among auditory streaming, dichotic fusion, and diotic fusion. Perception & Psychophysics, 32(2), 153–162.

213. Stevens, K. N. (2002). Toward a model for lexical access based on acoustic landmarks and distinctive features. Journal of the Acoustical Society of America, 111(4), 1872–1891.

214. Stoffregen, T. A., & Bardy, B. G. (2001). On specification and the senses. Behavioral and Brain Sciences, 24(2), 195–213.

215. Summerfield, C., & Egner, T. (2009). Expectation (and attention) in visual cognition. Trends in Cognitive Sciences, 13(9), 403–409.

216. Sussman, E. S. (2005). Integration and segregation in auditory scene analysis. Journal of the Acoustical Society of America, 117(3), 1285–1298.

217. Sussman, E. S. (2007). A new view on the MMN and attention debate – The role of context in processing auditory events. Journal of Psychophysiology, 21(3–4), 164–175.

218. Sussman, E. S., Gomes, H., Nousak, J. M. K., Ritter, W., & Vaughan, H. G. (1998). Feature conjunctions and auditory sensory memory. Brain Research, 793(1–2), 95–102.

219. Sussman, E. S., Horváth, J., Winkler, I., & Orr, M. (2007). The role of attention in the formation of auditory streams. Perception & Psychophysics, 69(1), 136–152.

220. Sussman, E. S., Ritter, W., & Vaughan, H. G. (1998a). Predictability of stimulus deviance and the mismatch negativity. NeuroReport, 9(18), 4167–4170.

221. Sussman, E. S., Ritter, W., & Vaughan, H. G. (1998b). Attention affects the organization of auditory input associated with the mismatch negativity system. Brain Research, 789(1), 130–138.

222. Sussman, E. S., Ritter, W., & Vaughan, H. G. (1999). An investigation of the auditory streaming effect using event-related brain potentials. Psychophysiology, 36(1), 22–34.

223. Sussman, E. S., Sheridan, K., Kreuzer, J., & Winkler, I. (2003). Representation of the standard: Stimulus context effects on the process generating the mismatch negativity component of event-related brain potentials. Psychophysiology, 40(3), 465–471.

224. Sussman, E. S., Winkler, I., Huotilainen, M., Ritter, W., & Näätänen, R. (2002). Topdown effects can modify the initially stimulus-driven auditory organization. Cognitive Brain Research, 13(3), 393–405.

225. Sussman, E. S., Winkler, I., Kreuzer, J., Saher, M., Näätänen, R., & Ritter, W. (2002). Temporal integration: Intentional sound discrimination does not modulate stimulus-driven processes in auditory event synthesis. Clinical Neurophysiology, 113(12), 1909–1920.

226. Sussman, E., Winkler, I., Ritter, W., Alho, K., & Naatanen, R. (1999). Temporal integration of auditory stimulus deviance as reflected by the mismatch negativity. Neuroscience Letters, 264(1–3), 161–164.

227. Sussman, E. S., Winkler, I., & Wang, W. J. (2003). MMN and attention: Competition for deviance detection. Psychophysiology, 40(3), 430–435.

228. Szalárdy, O., Bendixen, A., Bohm, T. M., Davies, L. A., Denham, S. L., & Winkler, I. } (2014). The effects of rhythm and melody on auditory stream segregation. Journal of the Acoustical Society of America, 135(3), 1392–1405.

229. Takegata, R., Brattico, E., Tervaniemi, M., Varyagina, O., Näätänen, R., & Winkler, I. (2005). Preattentive representation of feature conjunctions for concurrent spatially distributed auditory objects. Cognitive Brain Research, 25(1), 169–179.

230. Takegata, R., Huotilainen, M., Rinne, T., Näätänen, R., & Winkler, I. (2001). Changes in acoustic features and their conjunctions are processed by separate neuronal populations. NeuroReport, 12(3), 525–529.

231. Takegata, R., Paavilainen, P., Näätänen, R., & Winkler, I. (1999). Independent processing of changes in auditory single features and feature conjunctions in humans as indexed by the mismatch negativity. Neuroscience Letters, 266(2), 109–112.

232. Takegata, R., Paavilainen, P., Näätänen, R., & Winkler, I. (2001). Preattentive processing of spectral, temporal, and structural characteristics of acoustic regularities: A mismatch negativity study. Psychophysiology, 38(1), 92–98.

233. Teki, S., Chait, M., Kumar, S., Shamma, S., & Griffiths, T. D. (2013). Segregation of complex acoustic scenes based on temporal coherence. Elife, 2.

234. Teki, S., Chait, M., Kumar, S., von Kriegstein, K., & Griffiths, T. D. (2011). Brain bases for auditory stimulus-driven figure-ground segregation. Journal of Neuroscience, 31(1), 164–171.

235. Tervaniemi, M., Just, V., Koelsch, S., Widmann, A., & Schröger, E. (2005). Pitch discrimination accuracy in musicians vs nonmusicians: An event-related potential and behavioral study. Experimental Brain Research, 161(1), 1–10.

236. Tervaniemi, M., Maury, S., & Näätänen, R. (1994). Neural representations of abstract stimulus features in the human brain as reflected by the mismatch negativity. NeuroReport, 5(7), 844–846.

237. Tervaniemi, M., Rytkönen, M., Schröger, E., Ilmoniemi, R. J., & Näätänen, R. (2001). Superior formation of cortical memory traces for melodic patterns in musicians. Learning & Memory, 8(5), 295–300.

238. Thompson, W. F., Hall, M. D., & Pressing, J. (2001). Illusory conjunctions of pitch and duration in unfamiliar tone sequences. Journal of Experimental Psychology: Human Perception and Performance, 27(1), 128–140.

239. Tiitinen, H., May, P., Reinikainen, K., & Näätänen, R. (1994). Attentive novelty detection in humans is governed by pre-attentive sensory memory. Nature, 372(6501), 90–92.

240. Tishby, N., & Polani, D. (2011). Information theory of decisions and actions. In V. Cutsuridis, A. Hussain, & J. G. Taylor (Eds.), Perception-action cycle: Models, architectures, and hardware (pp. 601–636). New York, Dordrecht, Heidelberg, London: Springer.

241. Todd, J., Heathcote, A., Mullens, D., Whitson, L. R., Provost, A., & Winkler, I. (2014). What controls gain in gain control? Mismatch negativity (MMN), priors and system biases. Brain Topography, 27(4), 578–589.

242. Todd, J., Heathcote, A., Whitson, L. R., Mullens, D., Provost, A., & Winkler, I. (2014). Mismatch negativity (MMN) to pitch change is susceptible to order-dependent bias. Frontiers in Neuroscience, 8, 180.

243. Todd, J., Provost, A., & Cooper, G. (2011). Lasting first impressions: A conservative bias in automatic filters of the acoustic environment. Neuropsychologia, 49(12), 3399–3405.

244. Todd, J., Provost, A., Whitson, L. R., Cooper, G., & Heathcote, A. (2013). Not so primitive: Context-sensitive meta-learning about unattended sound sequences. Journal of Neurophysiology, 109(1), 99–105.

245. Tong, X., McBride, C., Zhang, J., Chung, K. K. H., Lee, C.-Y., Shuai, L., et al. (2014). Neural correlates of acoustic cues of English lexical stress in Cantonesespeaking children. Brain and Language, 138, 61–70.

246. Treisman, A. M. (1998). Feature binding, attention and object perception. Philosophical Transactions of the Royal Society of London, Series B: Biological Sciences, 353(1373), 1295–1306.

247. Treisman, A. M. (1993). The perception of features and objects. In A. Baddeley & L. Weiskrantz (Eds.), Attention: Selection, awareness, & control. A tribute to Donald Broadbent. Oxford: Clarendon Press.

248. Treisman, A. M., & Gelade, G. (1980). Feature-integration theory of attention. Cognitive Psychology, 12(1), 97–136.

249. van Norden, L. P. A. S. (1975). Temporal coherence in the perception of tone sequences. Eindhoven: Technical University.

250. van Petten, C., & Luka, B. J. (2012). Prediction during language comprehension: Benefits, costs, and ERP components. International Journal of Psychophysiology, 83(2), 176–190.

251. van Zuijen, T. L., Simoens, V. L., Paavilainen, P., Näätänen, R., & Tervaniemi, M. (2006). Implicit, intuitive, and explicit knowledge of abstract regularities in a sound sequence: An event-related brain potential study. Journal of Cognitive Neuroscience, 18(8), 1292–1303.

252. van Zuijen, T. L., Sussman, E., Winkler, I., Näätänen, R., & Tervaniemi, M. (2004). Grouping of sequential sounds – An event-related potential study comparing musicians and nonmusicians. Journal of Cognitive Neuroscience, 16(2), 331–338.

253. van Zuijen, T. L., Sussman, E., Winkler, I., Näätänen, R., & Tervaniemi, M. (2005). Auditory organization of sound sequences by a temporal or numerical regularity – A mismatch negativity study comparing musicians and nonmusicians. Cognitive Brain Research, 23(2–3), 270–276.

254. Versnel, H., Zwiers, M. P., & van Opstal, A. J. (2009). Spectrotemporal response properties of inferior colliculus neurons in alert monkey. Journal of Neuroscience, 29(31), 9725–9739.

255. Vliegen, J., & Oxenham, A. J. (1999). Sequential stream segregation in the absence of spectral cues. Journal of the Acoustical Society of America, 105(1), 339–346.

256. Wacongne, C., Changeux, J. P., & Dehaene, S. (2012). A neuronal model of predictive coding accounting for the mismatch negativity. Journal of Neuroscience, 32(11), 3665–3678.

257. Weise, A., Grimm, S., Müller, D., & Schröger, E. (2010). A temporal constraint for automatic deviance detection and object formation: A mismatch negativity study. Brain Research, 1331, 88–95.

258. Wessel, D. L. (1979). Timbre space as a musical control structure. Computer Music Journal, 3(7), 45–52.

259. Wetzel, N., Schröger, E., & Widmann, A. (2013). The dissociation between the P3a event-related potential and behavioral distraction. Psychophysiology, 50, 920–930.

260. Widmann, A., Kujala, T., Tervaniemi, M., Kujala, A., & Schröger, E. (2004). From symbols to sounds: Visual symbolic information activates sound representations. Psychophysiology, 41(5), 709–715.

261. Winkler, I. (2007). Interpreting the mismatch negativity. Journal of Psychophysiology, 21(3–4), 147–163.

262. Winkler, I., & Cowan, N. (2005). From sensory to long-term memory – Evidence from auditory memory reactivation studies. Experimental Psychology, 52(1), 3–20.

263. Winkler, I., Cowan, N., Csépe, V., Czigler, I., & Näätänen, R. (1996). Interactions between transient and long-term auditory memory as reflected by the mismatch negativity. Journal of Cognitive Neuroscience, 8(5), 403–415.

264. Winkler, I., & Czigler, I. (1998). Mismatch negativity: Deviance detection or the maintenance of the 'standard'. NeuroReport, 9(17), 3809–3813.

265. Winkler, I., & Czigler, I. (2012). Evidence from auditory and visual event-related potential (ERP) studies of deviance detection (MMN and vMMN) linking predictive coding theories and perceptual object representations. International

266. Journal of Psychophysiology, 83(2), 132–143.

267. Winkler, I. (2010). In search for auditory object representations. In I. Czigler & I.

268. Winkler (Eds.), Unconscious memory representations in perception: Processes and mechanisms in the brain (pp. 71–106). Amsterdam and Philadelphia: John Benjamins.

269. Winkler, I., Denham, S. L., Mill, R., Böhm, T. M., & Bendixen, A. (2012). Multistability in auditory stream segregation: A predictive coding view. Philosophical Transactions of the Royal Society B – Biological Sciences, 367(1591), 1001–1012.

270. Winkler, I., Denham, S. L., & Nelken, I. (2009). Modeling the auditory scene: Predictive regularity representations and perceptual objects. Trends in Cognitive Sciences, 13(12), 532–540.

271. Winkler, I., Karmos, G., & Näätänen, R. (1996). Adaptive modeling of the unattended acoustic environment reflected in the mismatch negativity event-related potential. Brain Research, 742(1–2), 239–252.

272. Winkler, I., Korzyukov, O., Gumenyuk, V., Cowan, N., Linkenkaer-Hansen, K., Ilmoniemi, R. J., et al. (2002). Temporary and longer term retention of acoustic information. Psychophysiology, 39(4), 530–534.

273. Winkler, I., Kujala, T., Tiitinen, H., Sivonen, P., Alku, P., Lehtokoski, A., et al. (1999). Brain responses reveal the learning of foreign language phonemes. Psychophysiology, 36(5), 638–642.

274. Winkler, I., Kushnerenko, E., Horváth, J., Ceponiene, R., Fellman, V., Huotilainen, M., et al. (2003). Newborn infants can organize the auditory world. Proceedings of the National academy of Sciences of the United States of America, 100(20), 11812–11815.

275. Winkler, I., Paavilainen, P., Alho, K., Reinikainen, K., Sams, M., & Näätänen, R. (1990). The effect of small variation of the frequent auditory stimulus on the eventrelated brain potential to the infrequent stimulus. Psychophysiology, 27(2), 228–235.

276. Winkler, I., Paavilainen, P., & Näätänen, R. (1992). Can echoic memory store two traces simultaneously? A study of event-related brain potentials. Psychophysiology, 29(3), 337–349.

277. Winkler, I., Reinikainen, K., & Näätänen, R. (1993). Event-related brain potentials reflect traces of echoic memory in humans. Perception & Psychophysics, 53(4), 443–449.

278. Winkler, I., & Schröger, E. (1995). Neural representation for the temporal structure of sound patterns. NeuroReport, 6(4), 690–694.

279. Winkler, I., Schröger, E., & Cowan, N. (2001). The role of large-scale memory organization in the mismatch negativity event-related brain potential. Journal of Cognitive Neuroscience, 13(1), 59–71.

280. Winkler, I., Sussman, E., Tervaniemi, M., Horváth, J., Ritter, W., & Näätänen, R. (2003). Preattentive auditory context effects. Cognitive Affective & Behavioral Neuroscience, 3(1), 57–77.

281. Winkler, I., Takegata, R., & Sussman, E. (2005). Event-related brain potentials reveal multiple stages in the perceptual organization of sound. Cognitive Brain Research, 25(1), 291–299.

282. Winkler, I., Teder-Sälejärvi, W. A., Horvath, J., Näätänen, R., & Sussman, E. (2003). Human auditory cortex tracks task-irrelevant sound sources. NeuroReport, 14(16), 2053–2056.

283. Winkler, I., van Zuijen, T. L., Sussman, E., Horváth, J., & Näätänen, R. (2006). Object representation in the human auditory system. European Journal of Neuroscience, 24(2), 625–634.

284. Woldorff, M. G., Hackley, S. A., & Hillyard, S. A. (1991). The effects of channelselective attention on the mismatch negativity wave elicited by deviant tones. Psychophysiology, 28(1), 30–42.

285.Woldorff, M. G., Hillyard, S. A., Gallen, C. C., Hampson, S. R., & Bloom, F. E. (1998). Magnetoencephalographic recordings demonstrate attentional modulation of vvmismatch-related neural activity in human auditory cortex. Psychophysiology, 35(3), 283–292.

286.Woods, D. L., & Alain, C. (2001). Conjoining three auditory features: An eventrelated brain potential study. Journal of Cognitive Neuroscience, 13(4), 492–509.

287.Woods, D. L., Alain, C., & Ogawa, K. H. (1998). Conjoining auditory and visual features during high-rate serial presentation: Processing and conjoining two features can be faster than processing one. Perception & Psychophysics, 60(2), 239–249.

288.Yabe, H., Tervaniemi, M., Sinkkonen, J., Huotilainen, M., Ilmoniemi, R. J., & Näätänen, R. (1998). Temporal window of integration of auditory information in the human brain. Psychophysiology, 35(5), 615–619.

289.Yabe, H., Winkler, I., Czigler, I., Koyama, S., Kakigi, R., Sutoh, T., et al. (2001). Organizing sound sequences in the human brain: The interplay of auditory streaming and temporal integration. Brain Research, 897(1–2), 222–227.

290.Yuille, A., & Kersten, D. (2006). Vision as Bayesian inference: Analysis by synthesis? Trends in Cognitive Sciences, 10(7), 301–308.

291.Zhuo, G., & Yu, X. (2011). Auditory feature binding and its hierarchical computational model. In H. Deng, D. Miao, J. Lei, & F. Wang (Eds.), Artificial intelligence and computational intelligence (pp. 332–338). Berlin, Heidelberg: Springer.

292.Zwislocki, J. J. (1969). Temporal summation of loudness – An analysis. Journal of the Acoustical Society of America, 46(2P2), 431–441.

CITATION

István Winkler, Erich Schröger, Auditory perceptual objects as generative models: Setting the stage for communication by sound, Brain and Language, Volume 148, September 2015, Pages 1-22, ISSN 0093-934X, http://dx.doi.org/ 10.1016/ j.bandl.2015.05.003.

Chapter 5

Ray Trace Modeling of Underwater Sound Propagation

Jens M. Hovem

Norwegian University of Science and Technology, Norway

INTRODUCTION

Modeling acoustic propagation conditions is an important issue in underwater acoustics and there exist several mathematical/numerical models based on different approaches. Some of the most used approaches are based on ray theory, modal expansion and wave number integration techniques. Ray acoustics and ray tracing techniques are the most intuitive and often the simplest means for modeling sound propagation in the sea. Ray acoustics is based on the assumption that sound propagates along rays that are normal to wave fronts, the surfaces of constant phase of the acoustic waves. When generated from a point source in a medium with constant sound speed, the wave fronts form surfaces that are concentric circles, and the sound follows straight line paths that radiate out from the sound source. If the speed of sound is not constant, the rays follow curved paths rather than straight ones. The computational technique known as ray tracing is a method used to calculate the trajectories of the ray paths of sound from the source.

Ray theory is derived from the wave equation when some simplifying assumptions are introduced and the method is essentially a high-frequency approximation. The method is sufficiently accurate for applications involving echo sounders, sonar, and communications systems for short and medium short distances. These devices normally use frequencies that satisfy the high frequency conditions. This article demonstrates that ray theory also can be successfully applied for much lower frequencies approaching the regime of seismic frequencies.

This article presents classical ray theory and demonstrates that ray theory gives a valuable insight and physical picture of how sound propagates in inhomogeneous media. However, ray theory has limitations and may not be valid for precise predictions of sound levels, especially in situations where refraction effects and focusing of sound are important. There exist corrective measures that can be used to improve classical ray theory, but these are not discussed in detail here. Recommended alternative readings include the books. [1-4] and the articles [5-6].

A number of realistic examples and cases are presented with the objective to describe some of the most important aspects of sound propagation in the oceans. This includes the effects of geographical and oceanographic seasonal changes and how the geoacoustic properties of the sea bottom may limit the propagation ranges, especially at low frequencies. The examples are based on experience from modeling sonar systems, underwater acoustic communication links and propagation of low frequency noise in the oceans. There exist a number of ray trace models, some are tuned to specific applications, and others are more general. In this chapter the applications and use of ray theory are illustrated by using Plane Ray, a ray tracing program developed by the author, for modeling underwater acoustic propagation with moderately range-varying bathymetry over layered bottom with a thin fluid sedimentary layer over a solid half with arbitrary geo-acoustic properties. However, the discussion is quite general and does not depend on the actual implementation of the theory.

THEORY OF RAY ACOUSTICS

The theory of ray acoustics can be found in most books and [1-4] will not be repeated here, but instead we follow a heuristic approach based on Snell's law, which is expressed by.

$$\xi = \frac{\cos\theta(z)}{c(z)} = \frac{\cos\theta_0}{c_0}.$$

(1)

Figure 1 shows a small segment of a ray path and the coordinate system. The segment has horizontal and vertical components dz and dr, respectively, and has the angle ϑ with the horizontal plane. When the speed of sound varies with depth the ray paths will bend and the rays propagate along curved paths. The radius of curvature R is defined as the ratio between an increment in the arc length and an increment in the angle

$$R = \frac{ds}{d\theta}.$$

(2)

Figure 1 shows that the radius of curvature is

$$R = \frac{1}{\sin\theta} \frac{dz}{d\theta}.$$

(3)

When the sound speed varies with depth the ray angle ϑ is a function of depth according to Snell's law. Taking the derivative of Eq. (3) with respect to gives the ray's radius of curvature at depth zexpressed as

$$R(z) = -\frac{c(z)}{\cos\theta(z)} \frac{1}{g(z)} = -\frac{1}{\xi g(z)}.$$

(4)

The ray parameter ξ is defined in Eq. (1) and $g(z)$ is the sound speed gradient.

$$g(z) = \frac{dc(z)}{dz}.$$

(5)

At any point in space, the ray curvature is therefore given by the ray parameter ξ and the local value of the sound speed gradient $g(z)$. The positive or negative sign of the gradient determines whether the sign of R is negative or positive, and thereby determines if the ray path curves downward or upward.

A ray with horizontal angle ϑin strikes a plane with inclination α, the reflected ray is changed to ϑ out.

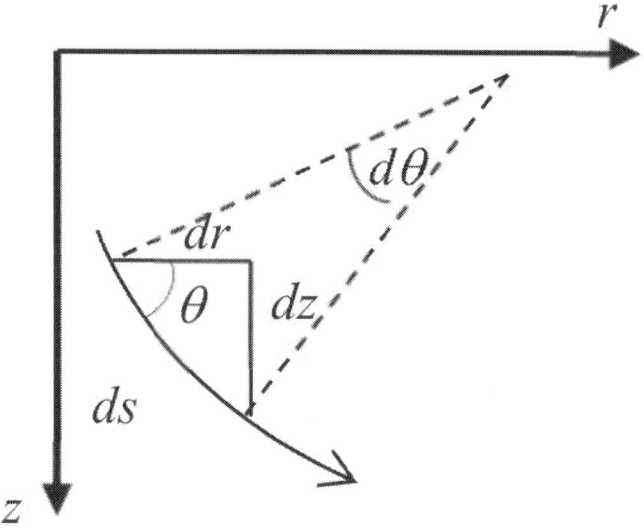

Figure 1. A small segment of a ray path in a isotropic medium with arc length ds.

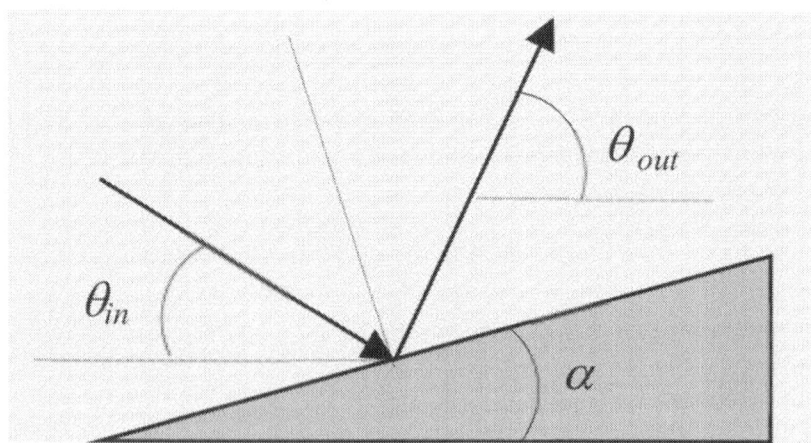

Figure 2. A ray with horizontal angle ϑin strikes a plane with inclination α, the reflected ray is changed to ϑ out.

The ray parameter is not constant when the bathymetry varies with range. The change in ray direction is illustrated in Figure 2 showing that after reflection the angle ϑin of an incoming ray is increased by twice the bottom inclination angle α.

$$\theta_{out} = \theta_{in} + 2\alpha .$$

(6)

Consequently, after the ray is reflected, its ray parameter must change from ξin to ξout, which is expressed as

$$\xi_{out} = \frac{\cos(\theta_{out})}{c} = \frac{\cos(\theta_{in} + 2\alpha)}{c}$$

$$= \xi_{in}\cos(2\alpha) \pm \frac{\sqrt{1 - \xi_{in}^2 c^2}}{c}\sin(2\alpha) .$$

(7)

The coordinates of a ray, starting with the angle ϑ_1 at the point (r_1, z_1), where the sound speed is c_1, as shown in Figure 3. For the coordinates of the running point at (r_2, z_2) along the ray path, the horizontal distance is

$$r_2 - r_1 = \int_{z_1}^{z_2} \frac{dz}{\tan\theta(z)} = \int_{z_1}^{z_2} \frac{\cos\theta(z)\,dz}{\sqrt{1 - \cos^2\theta(z)}} = \int_{z_1}^{z_2} \frac{\xi c(z)\,dz}{\sqrt{1 - \xi^2 c^2(z)}} .$$

(8)

The travel time between the two points is obtained by integrating the quantity $1/c$, the slowness, along the ray path:

$$\tau_2 - \tau_1 = \int_{z_1}^{z_2} \frac{ds}{c(s)} = \int_{z_1}^{z_2} \frac{dz}{c(z)\sin\theta(z)} = \int_{z_1}^{z_2} \frac{dz}{c(z)\sqrt{1 - \xi^2 c^2(z)}} .$$

(9)

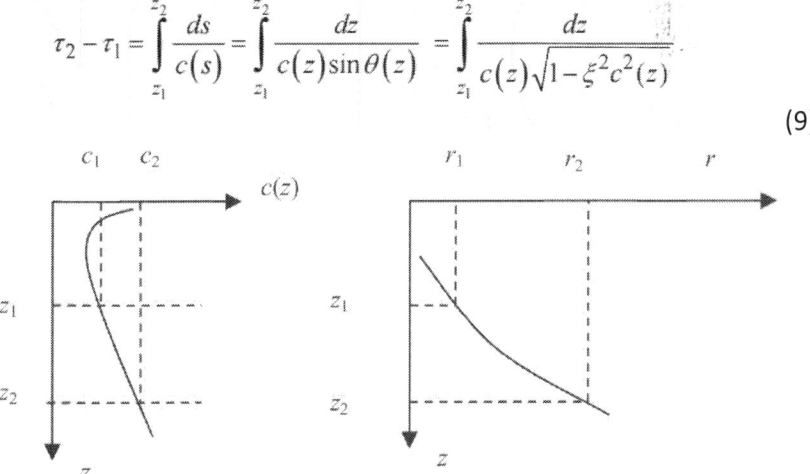

Figure 3. Left: The sound speed profile. Right: A portion of a ray traveling from point (r_1, z_1) to (r_2, z_2).

The acoustic intensity of a ray can, according to ray theory, be calculated using the principle that the power within a ray tube remains constant within that ray tube. This is illustrated in Figure 4 showing two rays with a vertical angle separation of $d\vartheta_0$ that define a ray tube centered on the initial angle ϑ_0. At a reference distance r_0 from the source, the intensity is I_0. Taking into consideration the cylindrical symmetry about the z axis, the power ΔP_0 within the narrow angle $d\vartheta_0$ is

$$\Delta P_0 = I_0 2\pi r_0^2 \cos\theta_0 d\theta_0 .$$

(10)

At horizontal distance r, the intensity is I. In terms of the perpendicular cross section dL of the ray tube, the power is

$$\Delta P = I 2\pi r dL .$$

(11)

Since the power in the ray tube does not change, we may equate Eq.(9) and eq. (10), and solve for the ratio of the intensities:

$$\frac{I}{I_0} = \frac{r_0^2}{r}\cos\theta_0 \left|\frac{d\theta_0}{dL}\right| .$$

(12)

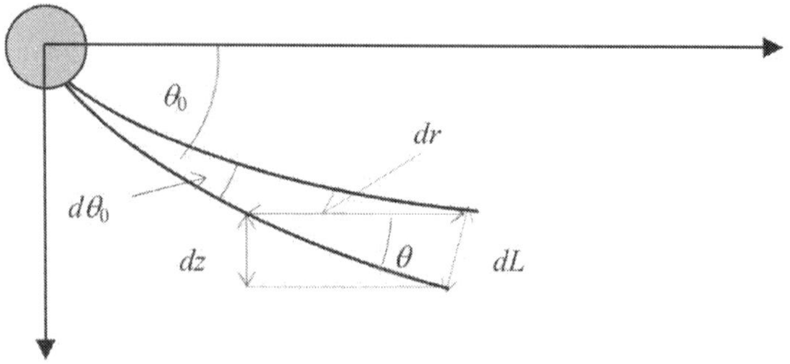

Figure 4. The principle of intensity calculations: energy radiated in a narrow tube remains inside the tube; r_0 represents a reference distance and ϑ_0 is the initial ray angle at the source; $d\theta_0$ is the initial angular separation between two rays; dr is the incremental range increase; ϑ is the angle at the field point; dz is the depth differential; and dL is the width of the ray tube.

Instead of using Eq.(11), it may be more convenient to use the vertical horizontal ray dr, which is

$$dr = \left| \frac{dL}{\sin \theta} \right|,$$

(13)

resulting in

$$\frac{I}{I_0} = \frac{r_0^2}{r} \left| \frac{\cos \theta_0}{\sin \theta} \frac{d\theta_0}{dr} \right| = \left(\frac{r_0^2}{r} \right) \left(\frac{c_0}{c} \right) \left| \frac{\cos \theta}{\sin \theta} \frac{d\theta_0}{dr} \right|.$$

(14)

The last expression in Eq.(13) is obtained by assuming that the ray parameter is constant and by using Snell's law. The absolute values are introduced to avoid problems with regard to the signs of the derivatives and of $\sin \vartheta$.

With respect to the reference distance r_0, the transmission loss TL is defined as

$$TL = -10 \log \left(I / I_0 \right).$$

(15)

By inserting Eq.(13) into Eq.(14) the transmission loss becomes

$$TL = 10 \log \left(r / r_0^2 \right) + 10 \log \left| \frac{dz}{d\theta_0} \right| + 10 \log \left(\frac{c}{c_0} \right).$$

(16)

The term c_0/c is close to unity in water and can be ignored in most cases. In this treatment the transmission loss includes only the geometric spreading loss. Therefore bottom and surface reflection losses and sea water absorption loss must be included separately.

The geometric transmission loss in Eq.(15) consists of two parts. The first term represents the horizontal spreading of the ray tube and results in a cylindrical spreading loss. The second and third terms represent the vertical spreading of the ray tube and are influenced by the depth gradient of the sound speed.

Eq.(13) predicts infinite intensity under either of two conditions: when ϑ = 0 or when $dr / d\vartheta_0 = 0$. The first condition signifies a turning point where the ray path becomes horizontal; the second condition occurs at points where an infinitesimal increase in the initial angle of the ray produces no change in the horizontal range traversed by the ray. The locus of all such points in space is called a caustic. In both cases there is focusing of energy by refraction and where classical ray theory incorrectly predicts infinite intensity. Caustics and turning points will be discussed further in section 8.2.

A RECIPE FOR TRACING OF RAYS

A simple receipt for a ray tracing algorithm is to divide the whole water column into a large number of layers, each with the same thickness Δz. Within each layer, the sound speed profile is approximated as linear so that, in the layer $zi < z < zi_{+1}$, the sound speed is taken to be

$$c\left(z\right) = c_i + g_i\left(z - z_i\right).$$

(17)

where ci is the speed at depth zi, and gi is the sound speed gradient in the layer. From Eq. (7) and Eq. (8) the range and travel time increments in the layer are given by

$$r_{i+1} - r_i = \frac{1}{\xi g_i}\left[\sqrt{1 - \xi^2 c^2(z_i)} - \sqrt{1 - \xi^2 c^2(z_{i+1})}\right],$$

(18)

and

$$\tau_{i+1} - \tau_i = \frac{1}{|g_i|}\ln\left(\frac{c(z_{i+1})}{c(z_i)}\frac{1 + \sqrt{1 - \xi^2 c^2(z_i)}}{1 + \sqrt{1 - \xi^2 c^2(z_{i+1})}}\right).$$

(19)

When $\xi^2 c^2(z_{i+1}) \geq 1$ the ray path turns at a depth between z_i and z_{i+1}, and Eq.s (17) and (18) must be replaced by the following expressions:

$$r_{i+1} - r_i = \frac{2}{\xi g_i} \sqrt{1 - \xi^2 c^2(z_i)} \,,$$

$$(20)$$

and

$$\tau_{i+1} - \tau_i = \frac{2}{|g_i|} \ln\left(\frac{1 + \sqrt{1 - \xi^2 c^2(z_i)}}{\xi c(z_i)} \right).$$

$$(21)$$

These equations give the trajectories and the travel times for any ray's path to the desired range. By applying Eqs. (13) and (14), the geometrical transmission loss is also determined. The simplicity of this method lies in the approximation of the sound speed profiles with straight-line segments and the ray path's subsequent decomposition into circular segments. The method's accuracy is determined by how well the linear fit matches the actual profile. In practice, the sound speed profile is often given as measured sound speeds at relatively few depth points. It is therefore advisable to use an interpolation scheme that is consistent with the usual behavior of the sound speed profile to increase the number of depth points to an acceptable high density.

The examples in this article are generated using the ray trace program PlaneRay that has been developed by the author [7-8]. However, any other ray programs with similar capabilities could have been use and the discussion is therefore valid for ray modeling in general. Other models frequently used and are the Bellhop model [9], and the models [10-11].
Figure 5 shows an example of ray modeling. The sound speed profile is shown at the left panel and the rays from a source at 50 m depth is shown in the right panel, which also shows the bathymetry and the thickness of the sediment layer over the solid half space.

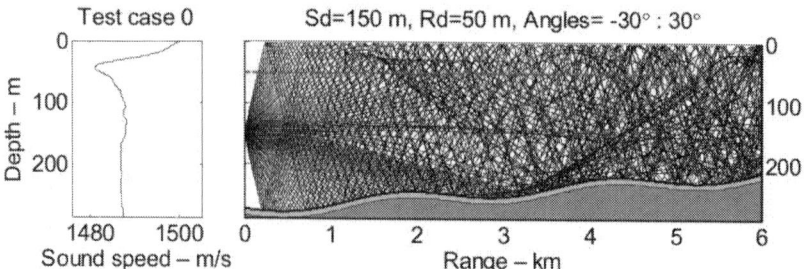

Figure 5. Sound speed profile and ray traces for a typical case. The source depth is 150 m and the red dotted line indicates a receiver line at a depth of 50 m. The initial angles of the rays at the source are from −30º to 30º.

EIGENRAY DETERMINATION

To calculate the acoustic field it is necessary to have an efficient and accurate algorithm for determination of eigenrays. An eigenray is defined as a ray that connects a source position with a receiver position. In most case with multipath propagation there are many eigenrays for a given source/receiver configuration, which means that finding all eigenrays is not a trivial task.

The PlaneRay model uses a unique sorting and interpolation routine for efficient determination of a large number of eigenrays in range dependent environments. This approach is described by the two plots in Figure 6, which displays the ray history as function of initial angle at the source. All facts and features of the acoustic fields such as the transmission loss, transfer function and time responses are derived from the ray traces and their history The two plots show the ranges and travel times to where the rays cross the receiver depth line (marked by the red dashed line in Figure 5). A particular ray may intersect the receiver depth line, at several ranges. For instance at the range of 2 km, there are 11 eigenrays and from Figure 6 the initial angles of these rays are approximately found to be 5.9°, 9.6°, 22°, 24° for the positive (down going) rays and−2.0°,−3.6, °− 7-4°− 15.0°− 17.0°− 25.0°,−27.0°, for the negative (up-going waves). However, the values found in this way are often not sufficiently accurate for the determination of the sound field. Further processing may therefore be required to obtain accurate results.

The graphs of Figure 6 are composed of independent points, but it is evident that the points are clustered in independent clusters or groups.

This property is used for sorting the points into branches of curves that represents different ray history. These branches are in most case relatively continuous and therefore amenable to interpolation. An additional advantage of this method is that the contribution of the various multipath arrivals can be evaluated separately, thereby enabling the user to study the structure of the field in detail.

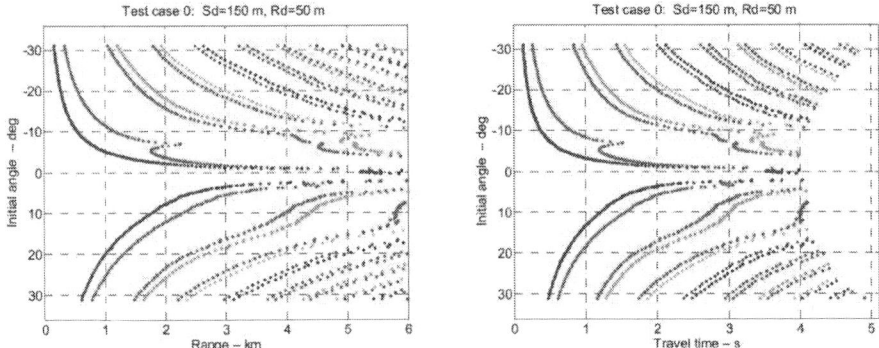

Figure 6. Ray history of the initial ray tracing in Figure 5 showing range (left) and travel time (right) to the receiver depth as function of initial angel at the source.

In most cases the eigenrays are determined by one simple interpolation yields values that are sufficiently accurate for most application, but the accuracy increases with increasing density of the initial angles at the cost of longer computation times.

Figure 7 shows examples of eigenrays traces with rays a receiver located at 2.5 km from the source for the scenario shown in Figure 5. To this receiver there are a total of 12 eigenrays, spanning the range of initial angles from -30° to 29°.

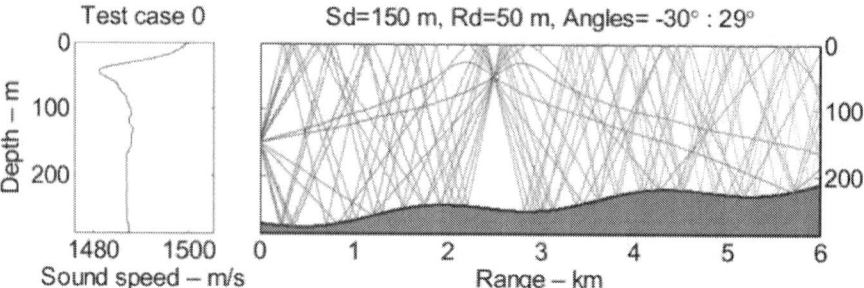

Figure 7. Eigenrays from a source at 150 m depth to a receiver at 50 m depth and distance of 2.5 km from the source.

ACOUSTIC ABSORPTION IN SEA WATER

Sound absorption is important for long range propagation especially at higher frequencies. The absorption increases with frequencies and is dependent on temperature, salinity, depth and the pH value of the water. There exists several expressions for acoustic absorption in sea water; one of the preferred options is the semi-empirical formulae by Francoise and Garrison [12]. Figure 8 shows sound absorption as function of frequency in sea water using this expression for the values given in the figure caption.

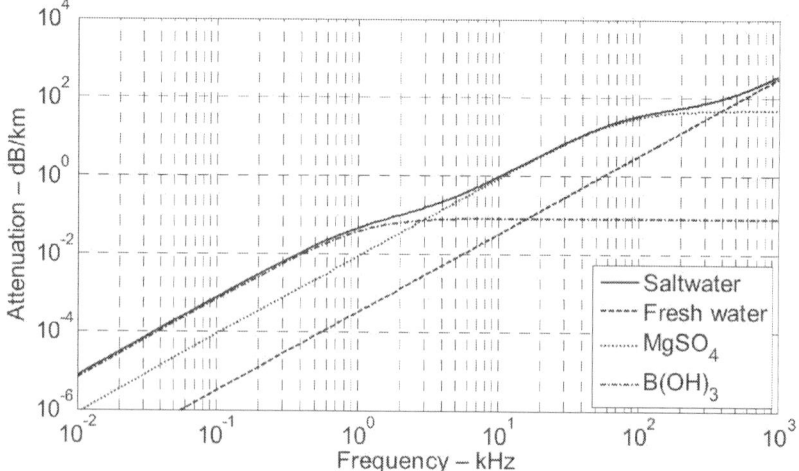

Figure 8. Acoustic absorption (dB/km) for fresh water and saltwater, plotted as a function of frequency (kHz) for water temperature of 10ºC, atmospheric pressure of one atmosphere (surface), salinity of 35 pro mille, and pH value of 7.8. The various contributions to the absorption are also indicated.

BOUNDARY CONDITIONS AT THE SURFACE AND BOTTOM INTERFACES

Ray tracing is greatly simplified when no rays are traced into the bottom, but stops at the water-bottom interface. This avoids tracing of multiple reflections in layered bottoms. Instead the boundary conditions at the sea surface and the bottom can be approximately satisfied by the use of plane wave reflection coefficient.

A simple and useful bottom model is assuming a fluid sedimentary layer over a homogeneous solid half space. The reflection coefficient of a bottom with this structure is

$$R_b = \frac{r_{01} + r_{12} \exp\left(-2i\gamma_{p1}D\right)}{1 + r_{01}r_{12} \exp\left(-2i\gamma_{p1}D\right)},$$

(22)

where γp_1 is the vertical wave number for sediment layer and D is the thickness of the sediment layer. The reflection coefficient between the water and the sediment layer, r_{01}, is given as

$$r_{01} = \frac{Z_{p1} - Z_{p0}}{Z_{p1} + Z_{p0}},$$

(23)

and r_{12} is the reflection coefficient between the sediment layer and the solid half space,

$$r_{12} = \frac{Z_{p2} \cos^2 2\theta_{s2} + Z_{s2} \sin^2 2\theta_{s2} - Z_{p1}}{Z_{p2} \cos^2 2\theta_{s2} + Z_{s2} \sin^2 2\theta_{s2} + Z_{p1}}.$$

(24)

In Eq (15) and (16) Zki is the acoustic impedance for the compressional ($k = p$) and shear ($k = s$) waves in water column ($i = 0$), sediment layer ($i = 1$) and solid half-space ($i = 2$), respectively. The grazing angle of the transmitted shear wave in the solid half-space is denoted ϑs_2.

Figure 9 shows an example of the bottom reflection loss as function of angle and frequency for a bottom with a sediment layer with the thickness $D = 2$ m with sound speed of 1700 m/s and density 1800 kg/m³ over a homogenous solid half space with compressional speed 3000 m/s, shear speed 500 m/s and density 2500 kg/m³. The wave attenuations are 0.5 dB/ wavelength. The critical angle changes from 60° at very low frequencies to about 28° at high frequencies, the two angles are given by the sound speed in the water and the two bottom sound speed of 3000 m/s and 1700 m/s. The small, but significant, reflection loss at lower angles is caused by shear wave conversion and bottom absorption In this case the attenuation is about 1 dB in the frequency band around 50 Hz to 100 Hz.

The reflection coefficient of a flat even sea surface is −1 for. For a sea surface with ocean waves there will be diffuse scattering to all other direction than the specular direction, which result in a reflection loss that in the first approximation can be modeled by the coherent rough surface reflection coefficient

$$R_{coh} = \exp\left[-2\left(\frac{2\pi}{\lambda}\sigma_h \sin\theta\right)^2\right].$$

(25)

In this expression ϑ is the grazing angle and σh is the rms. wave height and λ, is the acoustic wavelength, both in meters.

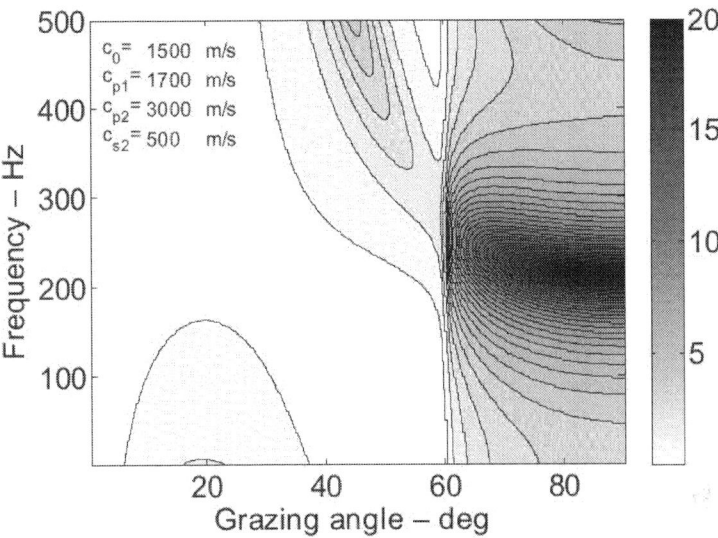

Figure 9. Bottom reflection loss (dB) as function of frequency and incident angle for a 2 m sediment layer over solid rock. The parameters are given in the text.

The reflection loss associated with reflection from a rough sea surface is

$$RL = -20\log 10\left(R_{coh}\right).$$

(26)

The same rough surface reflection coefficient may also be applied to a rough bottom.

Figure 10 shows the rough surface reflection loss as function of grazing angle, calculated for a wave height of 0.5 m and the frequencies of 50 Hz, 100 Hz, 200 Hz and 400 Hz.

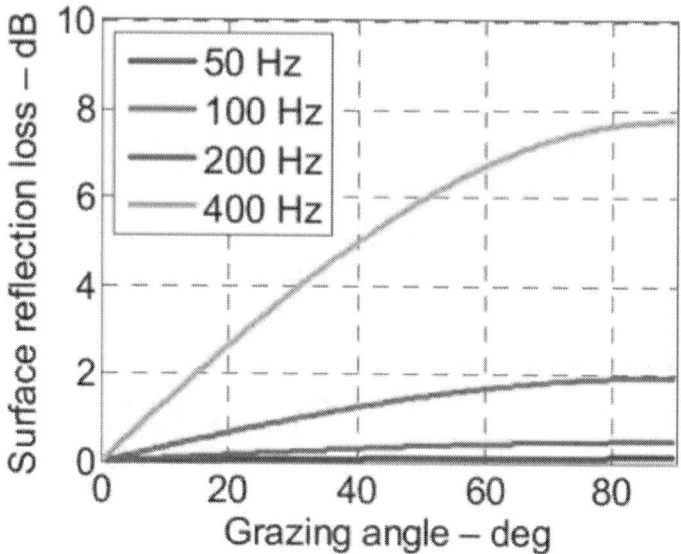

Figure 10. Reflection loss (dB) of rough surface with rms. wave height of 0.5 m as function of grazing angle, for the frequencies in the legend.

SYNTHESIZING THE FREQUENCY DOMAIN TRANSFER FUNCTION AND THE TIME RESPONSES

The total wave field at any receiving point is calculated in the frequency domain by coherent summation of all the eigenray contributions. The first step in the calculation is to determine the geometrical transmission loss of each of the multipath contributions by applying Eq. (13) and Eq.(14)to the sorted and interpolated range-angle values. The frequency domain transfer function and the transmission loss are obtained by adding the multipath contributions coherently in frequency domain taken into account the phase shifts associated the travel times from the interpolated history of the travel times. The frequency dependent acoustic absorption of sound in water is included at this point in the process. The transfer function $H(\omega, r)$ can be expressed as

$$H(\omega,r) = \sum_n A_n B_n(\omega) S_n(\omega) T_n \exp(i\omega\tau_n).$$

(27)

Eq. (26) expresses the transfer function $H(\omega, r)$ to a distance r from the source at the at angular frequency ω as a sum over the n eigenrays that are included in the synthesis. An is the geometrical spreading loss factor, defined as the square root of the expression in Eq. (13). Bn, and Sn, are the combined effects of all bottom reflections and surface reflections, respectively, Tn, is $-90\degree$ phase shift associated with caustics and turning points, and τn is the travel time.

The synthesis of the received signals is performed in the frequency domain by multiplying the frequency spectrum of the source signal with the transfer function of each of the eigenrays and summing the contributions. The time domain response is obtained after multiplication with the frequency function of a source signal followed an inverse Fourier transform of the product. This requires the choice of a source signal, a sampling frequency (fs) and a block length ($Nfft$) of the Fourier transform. The total duration of the time window (T_{max}) after Fourier transform is

$$T_{max} = \frac{N_{fft}}{f_s}.$$

(28)

It is important to select the values of $Nfft$ and fs such that Fourier time window, T_{max}, is larger than the actual length or duration of the signal. In reality the real time duration of the received signal is often not known in advanced and therefore the user may have to experiment with different values to find appropriate values for of $Nfft$ and fs.

Figure 11 shows an example where the transmission loss (in dB) as function of range has been calculated for the frequencies of 100 Hz and 200 Hz. The dashed black line indicates the geometrical spreading loss, which is added for comparison and given by,

$$TL_{geo}(r) = 10\log_{10}\left[r^2\left(1 + \frac{r^2}{r_t^2}\right)^{-1/2} \right].$$

(29)

This expression yields a transmission loss proportional to $20\log(r)$ when $r < rt$ and proportional to $10\log(r)$ for $r > rt$. This approximation to the geometrical transmission loss may be used for approximate calculations of transmission loss for flat bottom and simple sound speed profiles. In the case shown inFigure 11 rt is set equal to the water depth at source location, which in this case is 200 m.

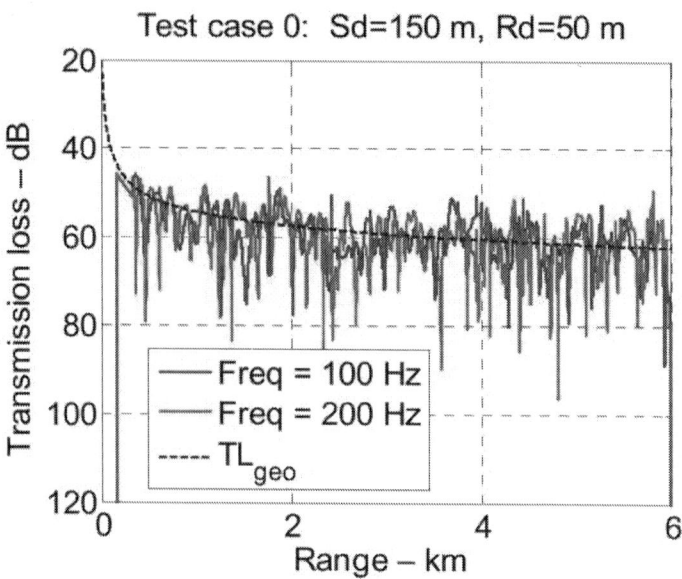

Figure 11. Transmission loss as function of range calculated for 100 Hz and 200 Hz The dashed black line is values of Eq.(28).

Figure 12 shows the synthesized time response at receivers spaced at 200 m separations in range up to 6 km. The sound speed and bathymetry is the same as in Figure 5 with the source at 150 m and all receivers at 50 m depth. The time scale is in reduced time to remove the gross transmission delay between the source and receiver. The reduced time is defined as

$$t_{red} = t_{real} - \frac{r}{c_{red}}.$$

(30)

In Eq. (29), *treal* and *tred* are the real and reduced times, respectively, *r* is range and *cred* is the reduction speed. The actual value of *cred* is not important as long as the chosen value results in a good display of the time responses.

In the example shown above, the time signal and calculated assuming a narrow band-limited source signal in the form of a Ricker pulse. An example of a Ricker pulse and its frequency spectrum are shown in Figure 13.

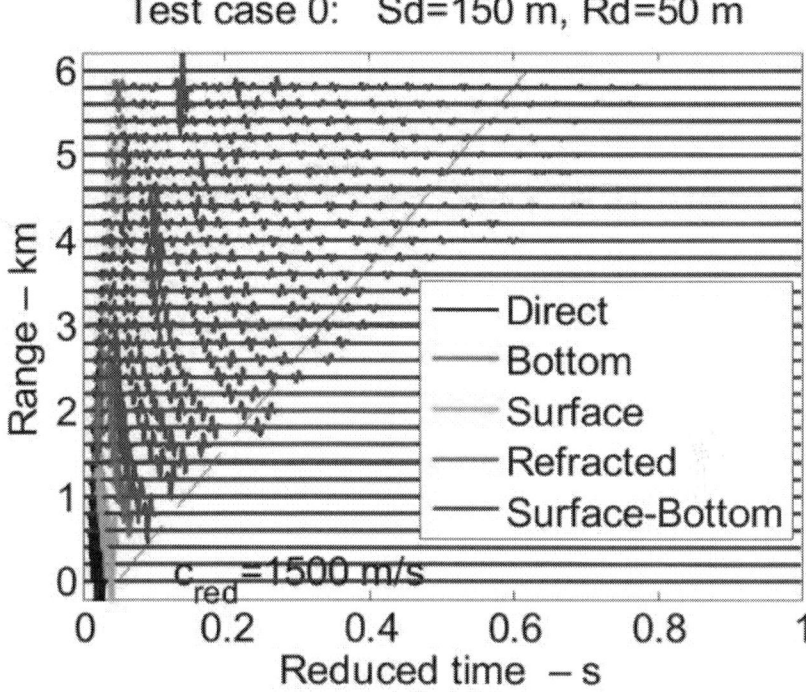

Figure 12. Received time signals as function of range and reduced time.

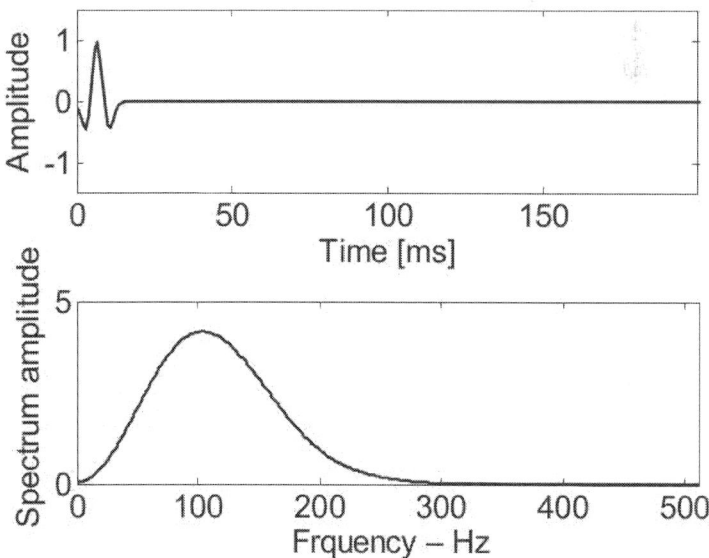

Figure 13. Ricker time pulse and frequency function the center frequency of 100 Hz

The time responses in Figure 12 are sorted according to the history of their eigenrays and color coded to allow for studying the various multipath contributions. This is particularly useful when dealing with transient signal and broad band signal, especially when knowledge of the multipath structure is important. In many such situations only the direct arrival or the refracted arrivals in the water column may carry the useful signals and all the other arrivals represent interference. In this case there are direct arrivals, followed by surface reflected and refracted arrivals at the turning points. Notice the high sound pressure values caused by the caustics at 3 km, 6, km and 7 km, which are apparent in both plots, this issue is discussed in section 8.2.

The red dotted line in Figure 12 represents an estimate of the duration of the cannel impulse response. This time duration is mainly given by the bottom reflection coefficient and the critical angle. Rays that propagate at angles closer to the horizontal plane than the critical angle experience almost no bottom reflection loss and may therefore propagate to long distances. Rays with steeper angles will experience higher reflection losses and die out more rapidly with range. Thus the time duration of the impulse is directly determined by the ratio of sound speeds in the water and the bottom as

$$t_{red} = \frac{r}{c_0}\left(\frac{1}{\cos\theta_{crit}} - 1\right) = r\left(\frac{c_b - c_0}{c_0^2}\right)$$

(31)

This estimate of the time duration of the channel impulse response assumes that the bottom is fluid, homogenous and flat, but the estimate may also be useful in other cases with moderately range dependent depth and with solid or layered bottom.

SPECIAL CONSIDERATIONS

Frequency of Applications
Ray tracing is a high frequency approximation to the solution of the wave equation and in principle more valid for high than for low frequency applications. However, high resolution prediction of higher frequency

acoustic fields is difficult both for numerical and physical reasons. Principally most important is the physical limitation caused by the fact that the sound speed and the environment are generally not known in sufficient detail. This can be illustrated by a simple example. Consider coherent communication using a frequency of 10 kHz with wavelength of 10 mm. The required accuracy in order to be correct at a distance of 1 km is that the sound speed is known and stable with a relative error less 10^{-5}, an impossible requirement to satisfy in practice regardless of the numerical accuracy of the computer model.

Caustics and Turning Points

As mentioned before, the locations where $dr/d\vartheta_0=0$ are called caustics where the ray phase is decreased by 90º and the where the intensity, according to ray theory, goes to infinity. In reality the intensity is high, but finite, and the basic ray theory breaks down at these points. There exists theories to amend and repair the defects of ray theory at these points [1, 2, 13], but that is not discussed here.

Figure 14 shows details of the field at a showing the rays with initial angles in the range of −6° to −1°. The scenario is the same as in of Figure 5, but for clarity the tracing of rays have been stopped after the first bottom reflection and the figure concentrates on the details the field at the caustic at 1760 m range for a ray with initial angle of −5.6°. Figure 15 shows the time responses for ranges in the interval from 1.6 km to 1.9 km. In this case, the source signal is a Ricker pulse with a peak frequency of 200 Hz. There is a first direct arrival (black color) at all ranges. From the range 1760 there is also a refracted arrival a little later than the direct, but with higher amplitude, in particular near the range of 1760 m. Notice the effect of the 90° phase shift for ranges beyond the caustic at 1760 m and that the amplitude at this range is considerable higher than at the other ranges.

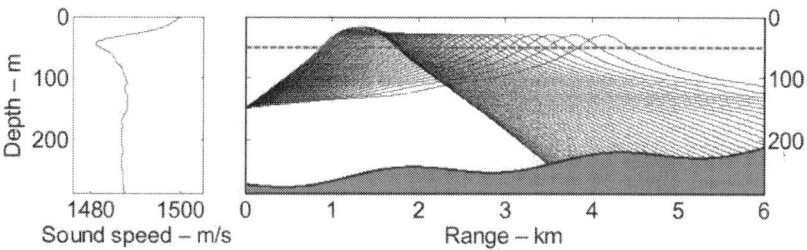

Figure 14. Rays through a caustic

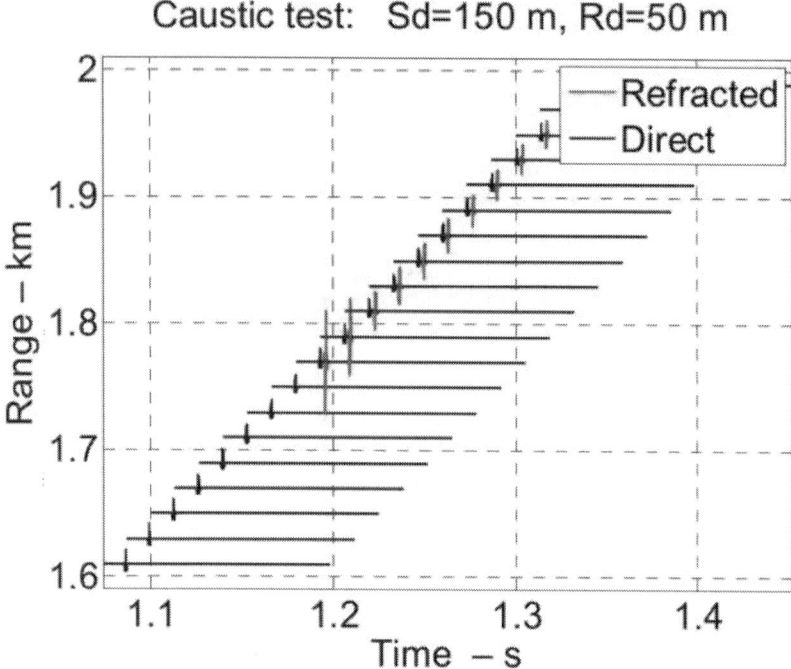

Figure 15. Time responses around the caustic at 1.76 km. The transmitted signal is a Ricker pulse with peak frequency of 200 Hz.

The Principle of Reciprocity and Its Validity in Ray Modeling

The principle of reciprocity is an important and useful property of linear acoustics and systems theory. The principle is very general and valid also in cases where the wave undergoes reflections at boundaries on its path from source to receiver [14]. The reciprocity principle is correctly represented in ray modeling, as can easily be understood from the eigenray plots of Figure 7. The eigenrays from a source position to the receiver position are the same as when source and receiver changes positions. The reflections at the bottom and at the sea surface are also symmetric in angles and consequently the acoustic fields are the same. However, it should be noted that the reciprocity principle applies to a point-to-point situation. This means that, for instance, that the development of the transmission loss as function source-receiver separation is generally not the same for the two directions.

The Validity of Using Plane Wave Reflection Coefficients

The accuracy of any ray model depends on the validity and limitation of ray theory and the implementation. A fundamental assumption of model is that the interactions with the boundaries are adequately described by plane wave reflection coefficient. In this section the validity of this assumption is investigated.

The general expression for the reflected field is given in text books, for instance in [13], over horizontal wave numbers k, as

$$\Phi_{ref}\left(r,z_r,\omega\right) = \frac{S(\omega)}{8\pi i}\int_0^\infty \Re(k)\frac{\exp\left(i\gamma\left|z_r + z_s\right|\right)}{\gamma}kH_0^{(1)}\left(kr\right)dk \ .$$

(32)

$\Phi ref(r, zr\ \omega)$ is the reflected field due to point source with frequency ω and source strength $S(\omega)$. $R(k)$ is the reflection coefficient, $H_0^1(kr)$ is the Hankel function of first kind, which represents a wave progressing in the positive r-direction. The horizontal wave number k and the vertical wave number γ are related to the sound speed, frequency and the angle ϑ by

$$k = \frac{\omega}{c}\cos\theta \ ,$$

$$\gamma = \frac{\omega}{c}\sin\theta \ .$$

(33)

Eq.(31) states that the field is given as an integral over all horizontal wave numbers, or as consequence of Eq.(32), integration over all angles both real and the imaginary.

Consider now the situation where R(k)=R is constant and independent of k or the angle. The integral in Eq.(31) becomes a standard integral and

$$\Phi_{ref}\left(r,z_r,\omega\right) = \frac{S(\omega)}{4\pi R}\Re\exp(ikR).$$

(34)

$$R = \left[r^2 + \left(z_s + z_r\right)^2\right].$$

(35)

According to Eq. (33) the reflected wave is the same as the outgoing spherical wave from the image of the source in the mirror position of the real source and modified by the constant reflection coefficientR. The situation with a constant reflection coefficient is valid for perfectly flat sea surface where the reflection coefficient is equal to -1 for all angles of incidence. Thus the reflection from a smooth sea surface is accurately described plane wave reflection coefficients.

In the general case, and for reflections from the bottom, the reflection coefficient R(k) is not constant and the integral can only be solved approximately or numerically. In order to obtain an approximation of the integral in Eq.(31), the Hankel function is expanded in a power series with the first terms being

$$H_0^1(kr) \approx \sqrt{\frac{2}{\pi kr}} \exp\left[i\left(kr - \frac{\pi}{4}\right)\right]\left[1 + \frac{1}{8ikr} + \cdots\right].$$

(36)

Restricting the integral of Eq.(31) to the first term yields

$$\Phi_{ref}(r, z_r, \omega) =$$

$$= \frac{S(\omega)}{4\pi} \frac{1}{\sqrt{2\pi r}} \int_{-\infty}^{\infty} \Re(k) \frac{\sqrt{k}}{\gamma} \exp\left[ikr + i\gamma(z_r + z_s)\right] dk.$$

(37)

The exponential in the integrand will normally be a rapid varying function and therefore the value of the integral will be small except when the phase term of Eq.(36) is nearly constant. The phase term ofEq.(36) is

$$\alpha = i\gamma(z_r + z_s) + ikr.$$

(38)

The stationary points are defined to the values of the horizontal wave number k where the derivative of the phase with respect to k is equal to zero, that is where $d\alpha/dk=0$, giving the stationary point as

$$r = \frac{(z + z_s)}{\tan(\theta_0)}.$$

(39)

The interpretation of this result is quite simple; the reflected wave field is equal to that of the image source multiplied with the reflection coefficient at the specular angleϑ_0.

There are however situations where this approximation is not sufficient in practice. This is discussed in [13] and in the following their results are cited without proof. The accuracy of the approximation depends on the source or receiver distance from the bottom interface. The result of the analysis is that the distance z from the bottom must satisfy

$$z \gg \frac{\lambda}{2\pi} \frac{\dfrac{\rho_b}{\rho_w}}{\sqrt{\left(\dfrac{c_b}{c_w}\right)^2 - 1}}.$$

(40)

With the water parameters of ρw = 1000 kg/m³ and cw=1500 m/s, and the bottom parameters of ρb = 1500 kg/m³ and cb=1700 m/s. Equation (39) requires than the distance from the bottom satisfy z>> 0.5 λ for the validity of using plane wave reflection coefficient at the bottom interface. A harder bottom with ρb = 1800 kg/m³ and cb=3000 m/s, gives the requirement that z>> 1.0 λ. Hence the condition for validity is somewhat easier to satisfy for a soft bottom than for a hard bottom.

Bench Marking Ray Modeling

The wave number integration model OASES [15] has been used to validate the accuracy and the limitation of the ray trace model using the simple case with constant water depth of 100 m and constant sound speed of 1500 m/s.

Figure 16 show the calculated transmission loss for the frequencies of 25 Hz, 50 Hz, 100 Hz and 200 Hz. The agreements between the results are very god for the higher frequency, but with some discrepancies for the lower frequencies, in particular for 25 Hz. The discrepancy is mainly a phase shift in the interference patterns of the two results, most pronounced for low frequencies and long ranges. This observation agrees with the theory outlined earlier. The seriousness of this discrepancy or errors may not very important in practice since the mean level is nearly the same as shown by the comparison with the OASES model.

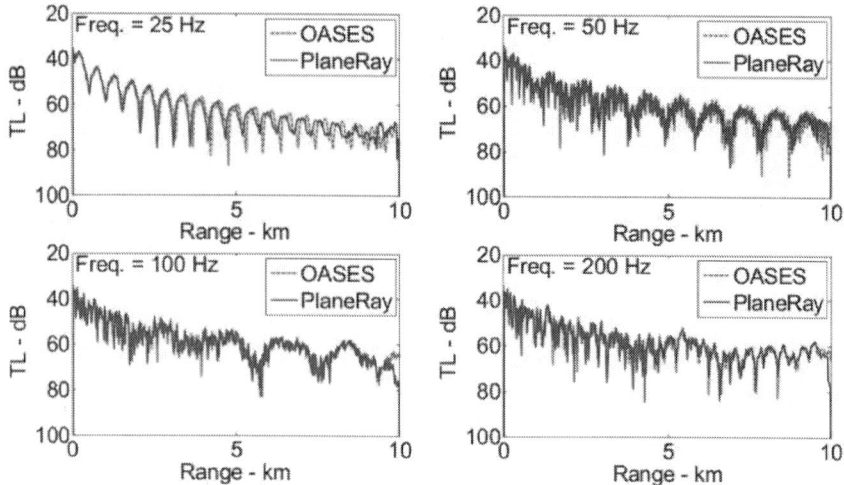

Figure 16. Comparison of the transmission loss as function of range for selected frequencies by PlaneRay (solid blue line) and OASES (dotted red line) for Pekeris' waveguide with a homogenous solid bottom with compressional wave speed of 3000 m/s and shear wave speed 500 m/s. Both wave attenuations have the values of 0.5 dB/wavelength.

CASE STUDIES

In the following we present two case studies that are relevant application of the modeling techniques descried in this article. The first if these is in connection with acoustic underwater communication and the transmission of digital information. In this case the multipath communication may be a significant problem causing intersymbol interference and significant degradation of reliability and performance. The second case is related to studies on the propagation of low frequency sound and the effect such noise may affect marine life, sea mammals and fish.

Seasonal Variations of Communication Links

In connection with a study of underwater acoustic communication the propagation over a 6 km track has been modeled for the various seasonal sound speed profiles.

The sound speed some months are shown in Figure 17. The sound speed profiles depend on the sea water temperature, the salinity and the depth. In the present case the sea water temperature variation with depth and the seasons is the main reason for changes in sound speed profile. During winters the surface water is cold and the sound speed is low, in the summer the surface water temperature and the sound speed is higher. The seasonal heating and cooling of the surface water propagates also to deeper depths, but with diminishing temperatures changes. At very large depths the water temperature is nearly the same at all seasons and the sound speed increases linearly and slowly with depth.

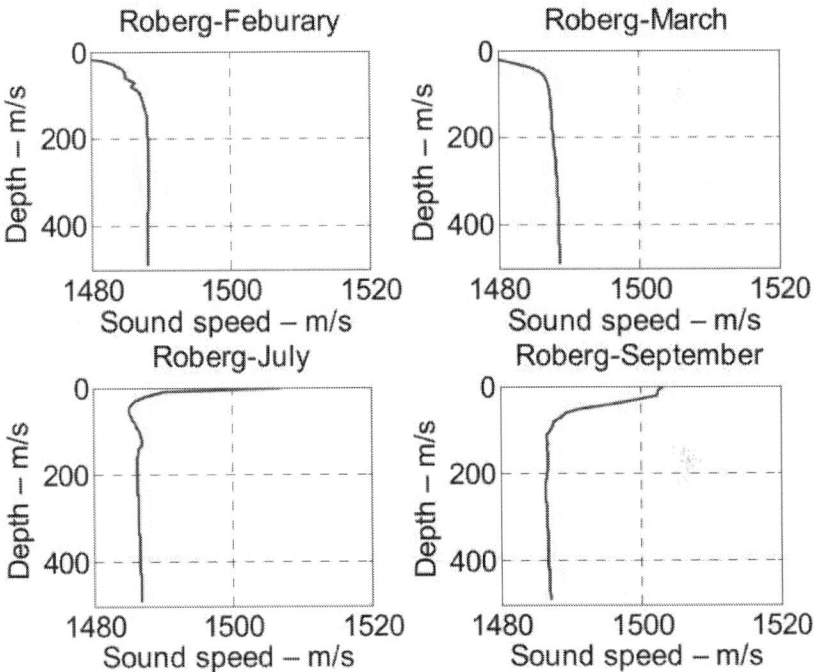

Figure 17. Sound speed profiles measured at specific dates for the months given in the figures

Figure 18 shows ray tracing results are for the same profiles as displayed in Figure 17. The purpose of the study was to investigate the possibility of communication to positions beyond a sea mount and to study the multipath arrival structure as function of range and depth.

There is a seamount with a peak at about 3 km from the transmitting station. In order to simplify the interpretation ray tracings in these plots have been terminated after 6 bottom reflections, but all rays are included in the calculation of the acoustic field, but rays with so many bottom reflections, or more, will in most case not be useful for data communication because of the reflection loss and reduced coherence.

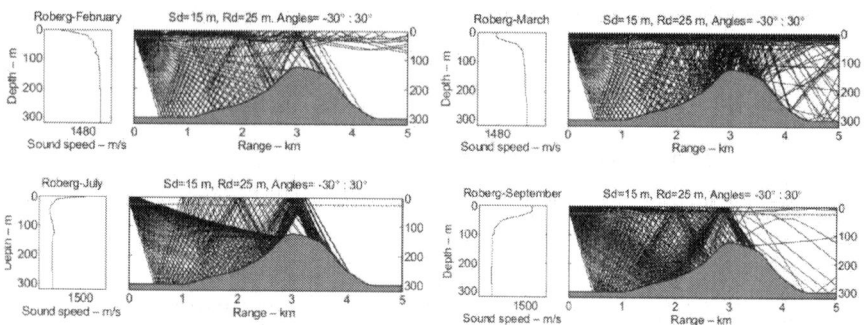

Figure 18. Ray tracing plots assuming a source depth of 15 meter for four monthly conditions at the Roberg test site. The sound speed profiles are the same as shown in Figure 17.

Figure 19 shows examples of received time responses at 25 m depth using a Ricker pulse as source signal. The different multipath contributions are color coded for clarity. At distances from the source over 1.5 km the first arrivals is follow paths surface reflected and upward refracted paths
Figure 20 shows the channel responses at a fixed range as function of depth down to 50 m. This figure shows the total response after adding all the individual multi path contributions. The plots demonstrate that the surface channel consists of deep refracted path and a number of paths reflected from the surface and deeper upwards refractions. The stability of these paths may be uncertain and subject to rapid changes in the environmental conditions near the surface due to temperature wind and current.

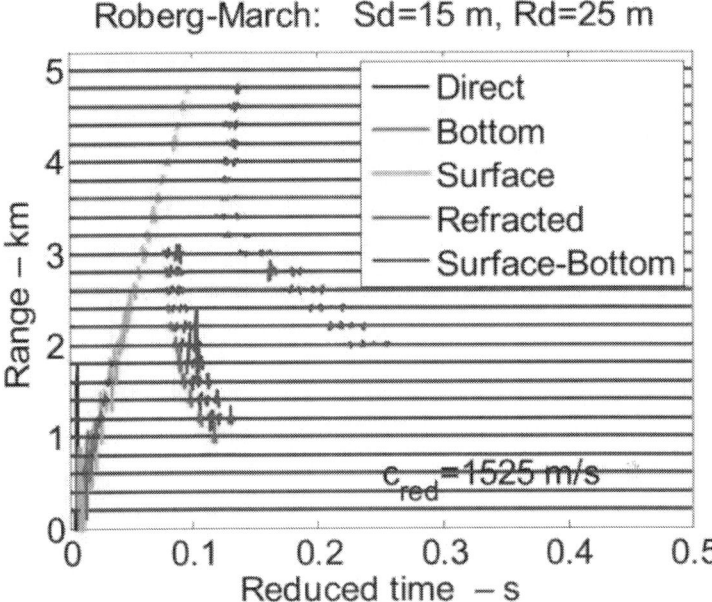

Figure 19. Time responses as function of range for receivers at depths of 25 m with a source at 15 m.

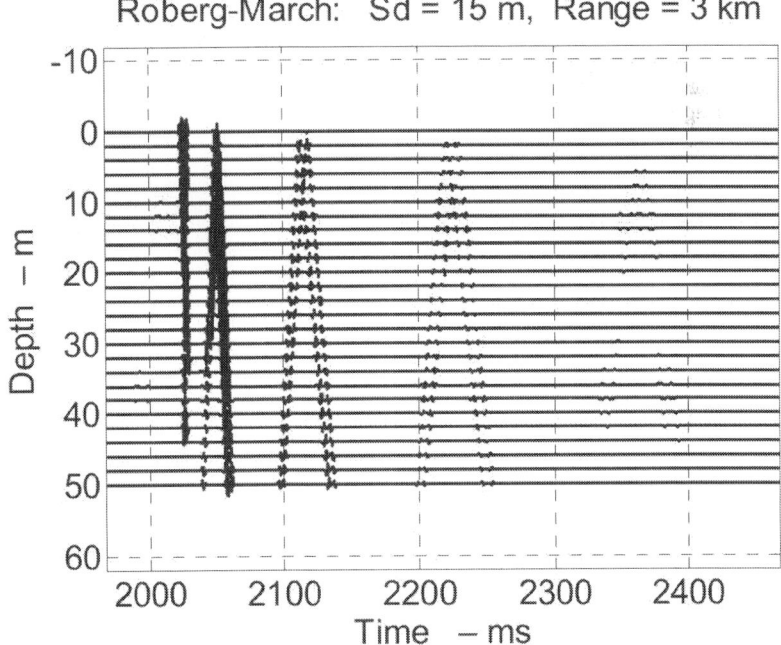

Figure 20. Time responses as function of receiver depth at a fixed horizontal distance of 3 km from a source at 15 m depth.

Seismic Noise Propagation

In many areas of the world anthropogenic noise often dominates over the natural ambient noise, especially in the low frequency band from approximately from 10 Hz and upwards to 1000 Hz, or more. This frequency band coincides approximately with the frequencies of perception of sea mammals and fish and may therefore be harmful to their natural activities, or even cause physical damages. An example is the case of the seismic exploration for oil and gas in certain areas where there is important commercial fishing interest. The propagation and distribution of acoustic noise depends the environmental conditions, in particular the oceanographic parameters, the topography of the seafloor and the acoustic properties of the bottom. In this section some of examples are presented to illustrate how the environment may affect the distribution of sound and noise. This study and discussion is also relevant for passive sonar applications to detect and track submerged vehicles and objects base emitted acoustic noise

The effects of bathymetric are illustrated in Figure 21 showing ray traces of upslope and downslope conditions for typical summer conditions at the Halten Bank in the Norwegian Sea. With downslope propagation there is a thinning the ray density with distance and upslope propagation gives a concentration of rays as the water depth diminishes.

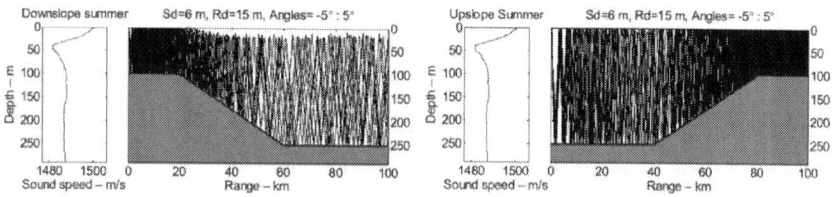

Figure 21. The effect of up and down sloping bottoms on the acoustic field distribution calculated for the typical summer condition in the month of July.

Figure 22 and Figure 23 show the calculated sound pressure level as function of range for the downslope and upslope propagation. The sound pulse from an airgun array is modeled as s a Ricker pulse with a peak pressure of 260 dB rel. 1µPa, centered on the frequency of 50 Hz, The horizontal dashed line is the assumed threshold value for fish reaction to sound. The bottom is modeled with a 2 m thick sedimentary layer over solid rock. The sound speed in the sediment layer is 1700 m/s and the density is 1800 kg/m³. The compressional sound speed in the rock is 3000 m/s, and density is 2500 kg/m³. The results in Figure 22 and Figure 23 are obtained under two conditions: (a) with a shear speed of 500 m/s, and (b) with no shear wave in the rock, i.e. the shear speed is zero. The absorptions are assumed to be 0.5 dB per wavelength for all the waves in the sediment layer and the rock. In the first case (a) the bottom reflection loss is as shown in Figure 9 with a significant low frequency reflection loss at angles lower than the critical angle caused by absorptions and conversion to shear wave in the bottom, which draws energy for the reflected wave. In the case of Figure 22 this results in a low-frequency and low-angle reflection loss of about 1 dB. For long ranges and many reflections this adds up to a significant total propagation loss. With no shear conversion the reflection loss is considerably reduced and the sound propagates easier to long ranges. The difference between the sound level at 50 Hz and 100 Hz is partly a result of increase attenuation at the higher frequency and partly that the source level in this case is higher for 50 Hz than for 100 Hz.

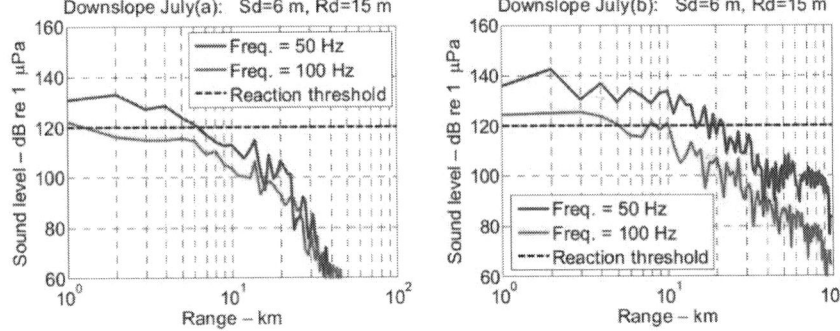

Figure 22. Sound pressure level as function of range for downslope propagation and July conditions. Left: With shear wave conversion (500 m/s). Right: No shear wave conversion.

Figure 24 and Figure 23 show similar results for downslope and upslope propagation for typical winter conditions represented by a sound speed profile measured in the month of February. For downslope conditions the sound level decrease rapidly with increasing depth and much more rapidly with shear wave conversion (Figure 24a) than without shear (Figure 24b). With upslope propagation (Figure 23) the sound levels are near independent of shear conversion except at the very long rages where the water depth becomes constant. The examples demonstrate that sound propagation in the ocean is strongly influenced by both by the oceanographic conditions and the geophysical properties of the bottom. Reliable prediction of acoustic propagation condition requires modeling tool that can that can handle both bottom and water properties.

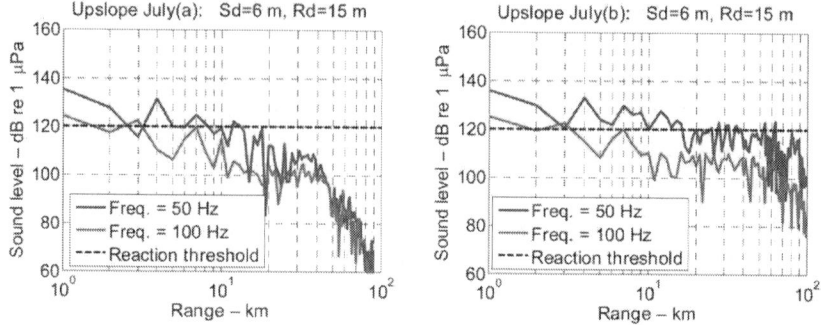

Figure 23. Sound pressure level as function of range for upslope propagation and July conditions. Left: With shear wave conversion (500 m/s). Right: No shear wave conversion.

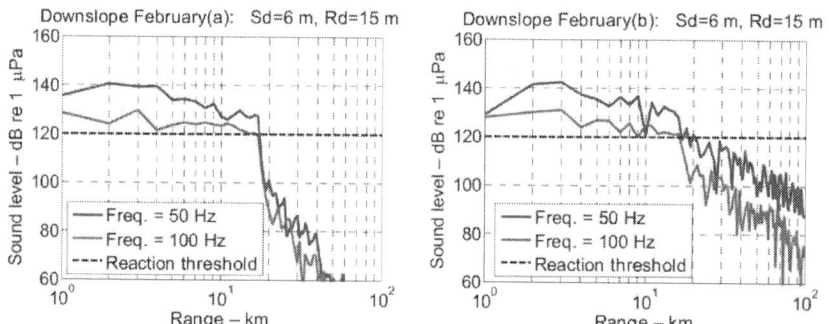

Figure 24. Sound pressure level as function of range for downslope propagation and February conditions. Left: With shear wave conversion (500 m/s). Right: No shear wave conversion.

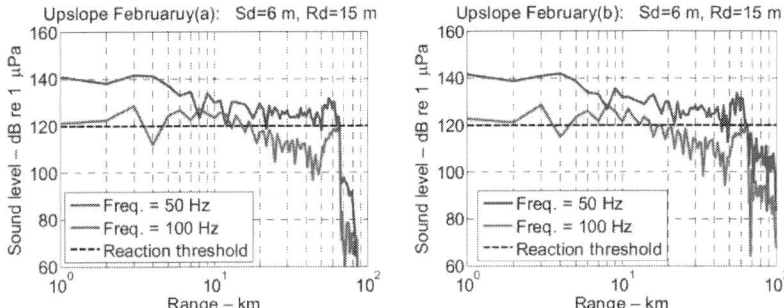

Figure 25. Sound pressure level as function of range for upslope propagation and February conditions. Left: With shear wave conversion (500 m/s). Right: No shear wave conversion.

SUMMARY

The article has outlined the theory of ray modeling and described how the theory can be applied to study acoustic wave propagation in the ocean. The complete acoustic fields are calculated by coherent addition of the contributions of a large number of eigenrays. In this method no rays are traced into the bottom, but the bottom interaction is modeled by plane wave reflection coefficients. Ray tracing is, by definition, frequency independent and therefore the ray trajectories through the water column are valid for all frequencies. Frequency dependency is introduced by reflections from the sea surface and the bottom, including loss associated with absorption and diffuse scattering of a rough ocean and bottom interfaces. Ray tracing is therefore a computational effective method for modeling broad of frequency band wave fields and for calculation of time responses.

Ray tracing is high-frequency approximation to the solution of the wave equation and the accuracy and validity at lower frequencies may be questioned, in particular the use of plane ray reflection coefficient to represent the bottom effects. This problem has been considered both theoretically and by simulations and comparison with more accurate model. The results of this study shows that source and receiver should be at a height above the bottom of at least half a wavelength, but there is no similar requirement to the distance from the sea surface. Less fundamental is the limitation of the numerical accuracy of the determination of the eigenrays, which is most serious in the calculation of the ray amplitude and the transmission loss. These inaccuracies are of more practical nature and can be reduced by refinements in the calculations.

Examples relevant for application in acoustic underwater communication and active sonar have been presented. The propagation of low frequency sound to large distances has been presented showing the effect of the bathymetry and the acoustic properties of the bottom. An important conclusion is the effect of bathymetry and the sound speed structure interacts and that accurate modeling of sound propagation requires information about the oceanography, the bathymetry and the geology of the bottom.

REFERENCES

1. C. B Officer, . (1958). Introduction to the theory of sound transmission. McGraw-Hill, New York City.
2. F. B. Jensen, W. A. Kuperman, M. B. Porter, and H. Schmidt, 2011Computational Acoustics, Second edition, Springer, New York.
3. L. E. Kinsler, A. R. Frey, A. B. Coppens, and J. V. Sanders, 2000Fundamentals of acoustics, 4th ed. New York: John Wiley & Sons.
4. J. M Hovem, . (2010). Marine Acoustics: The Physics of Sound in Underwater Environments. Peninsula Publishing, Los Altos, California, USA. 978-0-93214-665-6
5. E. K. Westwood, and P. J. Vidmar, 1987Eigenray finding and time series simulation in a layered-bottom oceanJ. Acoust. Soc. Am. 81, 912924
6. E. K. Westwood, and C. T. Tindle, 1987Shallow water time simulation using ray theory. J. Acoust. Soc. Am. 81, 17521761
7. Hovem Jens M. (2008). PlaneRay: An acoustic underwater propagation model based on ray tracing and plane wave reflection coefficients, in Theoretical and Computational Acoustics 2007, Edited by Michael Taroudakis and Panagiotis Papadakis, Published by the University of Crete, Greece, 273289 , (978-9-60897-854-2.
8. Hovem, Jens M. Ray trace modeling of underwater sound propagation. SINTEF Report A21539, 2011.11.23, 978-8-21404-997-8
9. M. B. Porter, 1991The Kraken normal mode programRep.SM-245. (Nato Undersea Research Centre), La Spezia, Italy. (1991).
10. L. Abrahamsson, 2003RAYLAB---a ray tracing program in underwater acoustics, Sci. Rep. FOIR--1047--SE, Division of Systems Technology, Swedish Defence Research Agency, Stockholm, Sweden.
11. S. Ivansson, 2006Stochastic ray-trace computations of transmission loss and reverberation in 3-D range-dependent environments, ECUA 2006, Carvoeiro, Portugal, 131136

12. R. E. Francois, and G. R. Garrison, 1982Sound absorption based on ocean measurements: Part II. Boric acid contribution and equation for total absorptionJ. Acoust. Soc. Am. 72(6), 1879-1890.
13. L. M. Brekhovskikh, and Yu. Lysanov. (2003Fundamentals of ocean acousticsrd ed. Springer-Verlag, New York City.
14. L. D. Landau, and E. M. Lifshitz, 1959Fluid Mechanics. Pergamon Press, Oxford UK.
15. H. Schmidt, 1987SAFARI: Seismo-acoustic fast field algorithm for range independent environments. User's guideSR-113, SACLANT Undersea Research Centre, La Spezia, Italy.

CITATION

Jens M. Hovem (2013). Ray Trace Modeling of Underwater Sound Propagation, Modeling and Measurement Methods for Acoustic Waves and for Acoustic Microdevices, Prof. Marco G. Beghi (Ed.), ISBN: 978-953-51-1189-4, InTech, DOI: 10.5772/55935.

Chapter 6

Acoustics in Optical Fiber

Abhilash Mandloi and Vivekanand Mishra

[1]Department of Electronics Engineering, S.V. National Institute of
Technology, Surat 395007, Gujarat, India

INTRODUCTION

Optical filters are the heart of optical networks; without the wavelength
selective device wavelength division multiplexing and dense wavelength
division multiplexing network will not exist. As the networks are
progressing towards closer wavelength spacing, performance
requirement for filters are becoming more demanding. Currently, the
popular filters include gratings, thin-film filters, and Fabry-Perot filters
and acoustoi optic tunable filters (AOTFs).

Acousto-optic (AO) effect in fibers has been studied to produce tunable
filters, gain flatteners, modulators, frequency shifters, and optical
switches reported. Most AO devices work on coupling from the
fundamental mode (LP_{11}) of light to a higher order asymmetrical (LP_{ll},
LP_{12}.... LP_{1n}) modes. Acousto-optics is defined as the discipline devoted to
the interactions between the acoustic waves and the light waves in a
material medium. Acoustic (vibrational) waves can be made to modulate,
deflect and focus light waves by causing a variation in the refractive
index. Acousto optic tunable filters are a promising technology for
dynamic gain equalization of optical fiber amplifiers [1]. By launching an
acoustic wave directly on the fiber, the device combines the merits of

fiber and AOTF devices namely the low insertion loss, low polarization dependence loss, wide tunability, fast tuning speed and ease of packaging. When a flexural acoustic wave is applied to a tapered single mode fiber, coupling takes place between the core mode and the cladding mode. The coupled energy in the cladding mode is essentially absorbed by the fiber jacket as reported so that the device is a notch filter. It means the centre frequency and the rejection efficiency can be tuned by adjustment of the frequency and voltage being applied. Varying the amplitudes and frequency of a RF generator can change the spectral profile of these filters.

To improve the rejection efficiency of the filters, the thickness of the fiber can be reduced. This is achieved through the heating and the acid-etching method. In the heating method, the ratio of cladding to core size is maintained while in the acid etching-method, the ratio between the cladding and core can be changed.

Acousto-Optic Tunable Filter

Device Design
The fiber used in our experiment is a Corning SMF-28, standard telecommunication single mode fiber. A region of SMF is etched by dipping the fiber in a hydrofluoric acid solution, which has a concentration of 40%. Etching rate controls the thickness of the SMF and the diameter reduction is observed using a CCD camera.

When the optical signal enters the fiber and interacts with the acoustic energy in a jacket stripped segment of the fiber, the core mode of the light is converted to a higher order cladding mode producing a notch filter like characteristics in the transmission spectrum. Core mode converting to various cladding modes will produce a few notch filters, with each having its peak notch at a separate wavelength [2-4]. A vibrating PZT transducer driven by a RF generator produces the acoustic energy as stated by Yun, Hwang and Kim, (1996). The acoustic energy is further amplified and concentrated to the fiber by a machined aluminium horn.

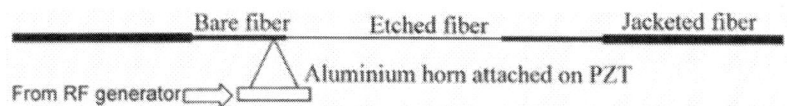

Figure 1: The setup to study AO interaction inside a fiber.

Horn Design

An acoustic horn functions to transfer and amplify the surface acoustic wave to the fiber. All horns made were conical in shape, where the tip is narrow and the base is broad as described by Lee, Kim Hwang and Yun, (2003). All the horns fabricated for the AOTF experiment have a ratio of length to outer diameter ratio of 2. Length is defined as the length from the tip of the horn to the base of the horn. Horns taken are 1cm in length. Outer diameter of the horn is defined at the diameter at the base. The inside of the horns are made hollow. When the horn is made considerably small, the frequency dependence on the acoustic generator is low. In the experiments done, no transduction is observed when the fiber is not etched. Potential problems can be attributed to the size of the transducer and the adhesive used to bond the tip of the horn to the fiber. Solder acts as a strong, metallic, thermally stable, and acoustically transmitting joint. In these experiments however glue was chosen as the bonding agent. Of particular interest will be the horn tip size. Acoustic impedance at the horn tip is given by:

$$Z_r = c_a v_a A_r$$

(1)

where c_a is the longitudinal velocity inside aluminium, v_a is the density aluminium and A_r is the cross section of the horn tip. Acoustic impedance at the bond junction along the fiber is given by:

$$Z_s = 2c_s v_s A_s$$

(2)

where c_s is the longitudinal velocity inside silica, v_s is the density silica and A_s is the cross section of fiber. Acoustic impedance inside the fiber is counted twice because of bidirectional acoustic movement along the fiber. Optimum transduction occurs when $Z_r = Z_s$ and since acoustic impedance of silica is almost matching that of aluminium, according to engan et. al maximum acoustic wave transfer occurs when horn tip diameter is almost matching that of the fiber.

Tuning of Peak Wavelength

By driving a piezoelectric (PZT) device at an ultrasonic frequency the periodic perturbations can be created inside the fiber. In a phase-matched condition, where the momentum and energy conservation requirement (LB =^) are met, the resonant frequency of an acoustic wave according to Birks, Russel and Culverhouse (1992) is given by

$$ f = \frac{\pi b C_{ext}}{L_B^2} = \frac{\pi b C_{ext}}{\Lambda^2} $$

(3)

where b is the radius of the fiber, C_{ext} is the speed of fundamental acoustic mode, which for silica is 5760 ms^{-1}, $\wedge$ is the period of the microbend[1] - .

Assuming a phase-matched condition, the frequency needed to transfer the modes from core to cladding mode for various thickness of the fiber is given in Fig 2 and Fig 3. As the fiber diameter is reduced, the values of $df/d\lambda$ get smaller. For unetched fibres, the frequencies used to create the micro bends and thus, convert the modes are from 1.75 MHz to 2.25 MHz. For thin diameter fibers (20 μ m, 30 μ m, 40 μ m), the frequencies are from 800 kHz to 1.1 MHz.

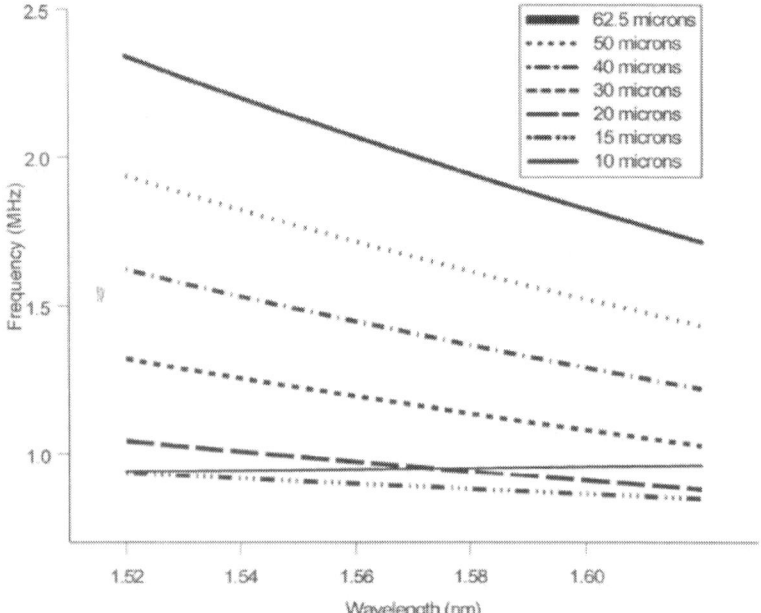

Figure 2: Calculated RF frequency to convert the LP01 mode to LP11 mode plotted against wavelength (for various thickness of fibre diameter).

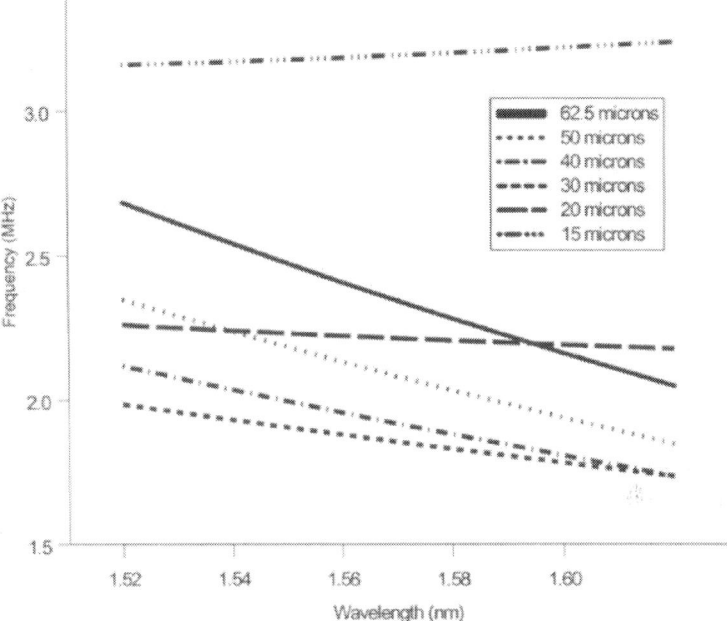

Figure 3: Calculated RF frequency to convert the LP0l mode to LP12 mode plotted against wavelength (for various thickness of fibre diameter).

Frequency from the RF generator can be used to control the peak wavelength tuning of the notch filters (Fig.4). The fiber used in the experiment has a diameter of 30 μm, and length of 17 cm. Higher frequencies of the RF generator will blue shift the peak wavelength of the filter [3-7]. The tuning range of the filter is slightly less than 300 nrn. From Eq.1.3, we deduce that, micro bend's period is inversely proportional to the frequency of the RF generator. For a larger value of period, the filter's peak is red shifted. Thin fibers have lower period values, thus etching the fibres will blue shift the peak wavelength of the notch filters. Frequency used to tune the peak wavelength as in for thin fibres is from 800 kHz to 1.1 MHz, which is in excellent agreement with the theory as in Fig.3 and Fig.4.

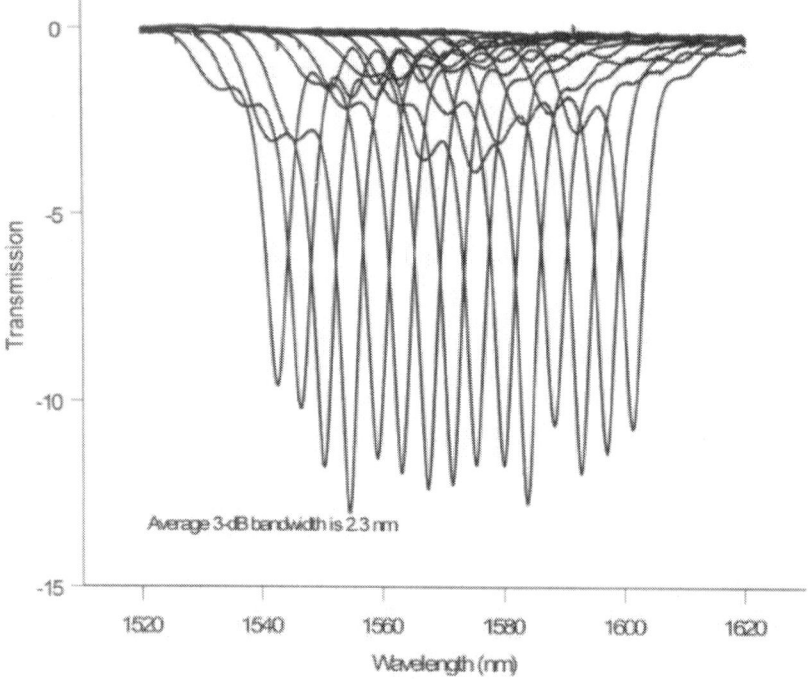

Figure 4: Measured peak wavelength tuning of the filter by changing the RF frequency. Frequency used is from 970 kHz to 1045 kHz. The fiber used has a thickness of 30 μm and length of 17 cm.

Tuning of Attenuation Depth

The RF generator's V_{p-p} level will be used to control the attenuation depth of the filter. V_{p-p} level is actually referring to the acoustic power transferred to the fiber. Increasing the V_{p-p} level will generally increase the bottom level of the filter as seen from Fig. 5. However in some cases increasing the V_{p-p} level will only distort the shape of the filter without increasing the notch's depth. For this strong over-coupled phenomenon, side lobes of the filter is actually increasing. One way to eliminate the problem is by limiting the interaction length of light inside the etched region. Here the power means RF generator's Vp-p level which will be used to control the attenuation depth of the filter[8-10]. Vp-p level is actually referring to the acoustic power transferred to the fiber. Increasing the Vp-p level will generally increase the bottom level of the filter as seen from Fig. 5. Here acoustic power supplied to PZT is 1.6 W to allow mode conversion.

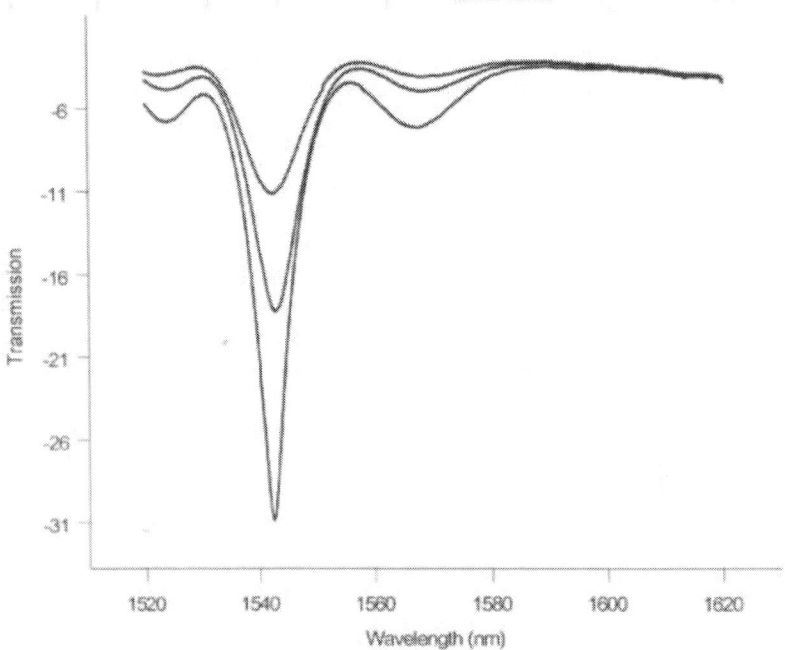

Figure 5: Measured attenuation variation of filter when the power of RF generator (V_{p-p}) is increased.

An effort to reduce the acoustic power fed into the fiber is by reducing the thickness of the fiber. The minimum acoustic power required by the device to operate or to allow mode conversion, is given by

$$P = 2\pi^3 \rho v_g (fR)^2 (u_t)^2$$

(4)

where ρ is the mass density of the fiber (ρ =2200kg/m3 for fused silica), V_g is the group velocity of the wave and R is is curvature of fiber, and u_t is the transverse acoustic amplitude which is given by:

$$u_t = \frac{\pi}{2} L_B \frac{a}{L} \frac{1}{0.908}$$

(5)

where L is referring to the interaction length of acoustic and light inside the fiber and LB is the optical beat length. Fig. 6 shows the calculated power required for mode conversion is lower for etched fibers. When the fiber is unetched the power required will be 287 mW. For a 20 µ m fiber, the power required for conversion is only 1.17 mW.

Experimentally, as seen from Fig. 7 for a 37 μm thick fiber, the acoustic power supplied to PZT is 1.6 W to allow mode conversion. Mode conversion was confirmed using far-field radiation pattern as reported by Doma, and Blake (1992). However, in a *26 μ* m fiber, power requirement for mode conversion is reduced to a mere value of 42 mW. The difference in the power reduction with the calculated value, suggests that the loss at the point of contact is high[10]. It is believed that the horn design is still not optimized; nevertheless, this transduction is sufficient to demonstrate conversion between two modes. Typically, only the lowest order flexural acoustic mode should be made to travel inside the fiber, and this can be achieved by ensuring the horn tip's thickness is matching that of the fiber.

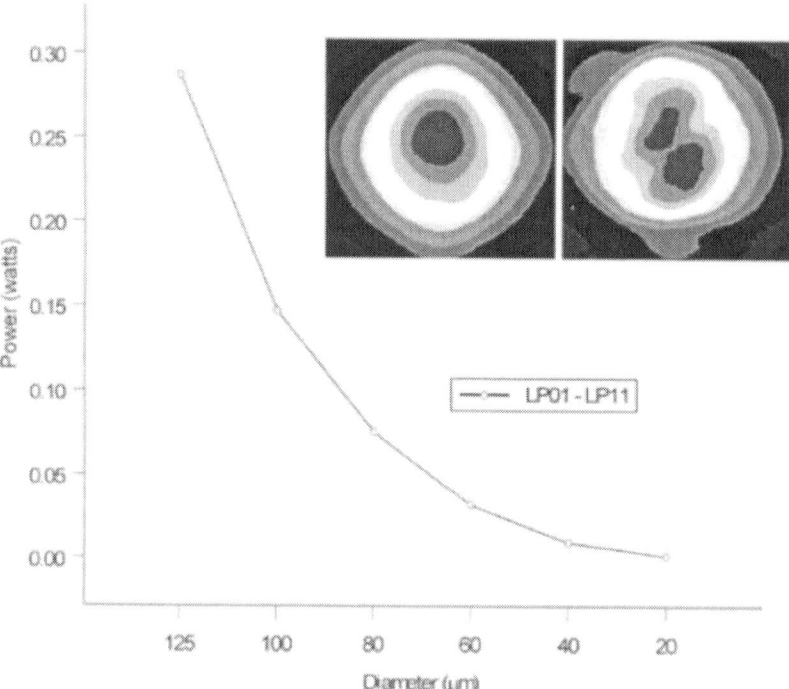

Figure 6: Calculated acoustic power required to allow mode conversion. Interaction length was set to 13 cm. Inset: Far field radiation pattern of modes involved in conversion. Left- LP01, Right- LP11.

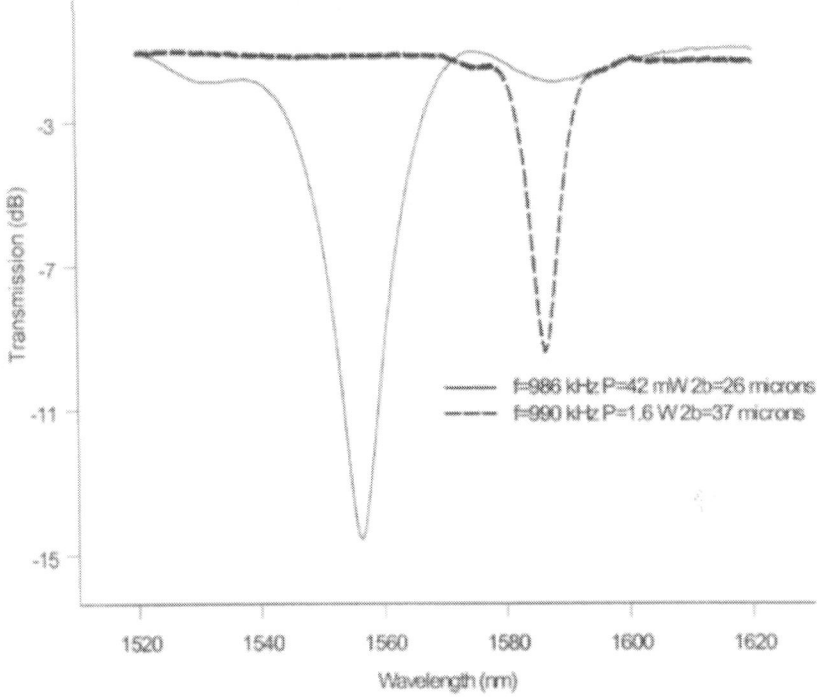

Figure 7: Measured transmission spectrum when fiber is etched. Significant reduction in acoustic power is observed.

Tuning of 3-Db Bandwidth

The 3-dB bandwidth of the notch filter is given by the equation below as reported by D. Ostling, H.E. Engan (1995):

$$\Delta\lambda = \frac{0.8}{L}\left[\frac{\partial L_B(\lambda)}{\partial\lambda}\right]^{-1}\left[L_B(\lambda)\right]^2$$

(6)

Where λ is the wavelength of the light, L is the length of the coupling interaction, and *LB* is the optical beat length[11-14]. For a broadband filter, a short coupling length, a long beat length and small beat length dispersion is required. Without making the device short, only by etching the fiber to that thickness a broad filter can be obtained as reported. However this bandwidth is not tuneable and so is not suitable for spectral shaping. In this section, a similar achievement by only using a SMF to tune the 3-dB bandwidth of the filter is demonstrated. In this device, the notch filter's attenuation, peak wavelength tuning and 3-dB bandwidth can be simultaneously controlled in a single device.

To achieve this, a tunable acoustic absorber is added to the original AO setup as shown in Fig. 8. By moving the acoustic absorber along the etched region of the fiber, the interaction of light inside the acoustic region can be controlled. From Eq.6, we know that by controlling the coupling interaction length the 3-dB bandwidth of the filter can be controlled [15]. A strong acoustic absorbing material such as cotton or polystyrene can be used as the acoustic absorber [16]. The absorbing material functions to ensure no surface acoustic wave beyond the absorber's position are present. Since the interaction length of light inside the acoustic region can be controlled, over-coupling phenomenon can be monitored, to reduce the effects of undesirable side lobes. Broad filters require higher power to operate when the attenuation level is maintained the same as a narrow filter.

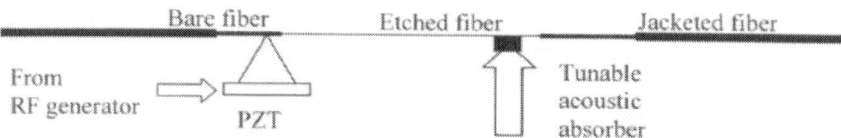

Figure 8: Setup to study the bandwidth variation using the AO interaction inside fiber.

From Fig. 9, the narrowest filter with a 3-dB bandwidth of 13 nm is obtained when maintaining the interaction length at 14 cm while the broadest filter with a 3-dB bandwidth of 28 nm is obtained when the interaction length is reduced to 7 cm. For an interaction length of 11.5 cm the spectral width is 16 nm and for 9 cm the spectral width will be 21 nm. The frequency used for wavelength tuning was from 960 kHz to 995 kHz and was sufficient to cover a wavelength span of 100 nm (1520 om-1620 nm).

For the broadest filter, the RF generator's V_{p-p} needed to generate coupling between the modes, seem to be the highest at 14 V. Meanwhile, for the narrowest filter, the V_{p-p} needed is only 6 V. Thus, we need a higher V_{p-p} to generate filters for shorter interaction length of light inside the grating region [17]. The introduction of tuneable acoustic absorber will change the strain dependency on the device. To limit the strain change introduced in the device only the tip of the absorber is allowed to touch the fiber, in our case, the resonant frequency change corresponding to the strain change was maintained at +/- 0.7kHz.

Throughout the experiment the total IL was maintained less than 0.1 dB and the PDL was less than 0.4 dB.

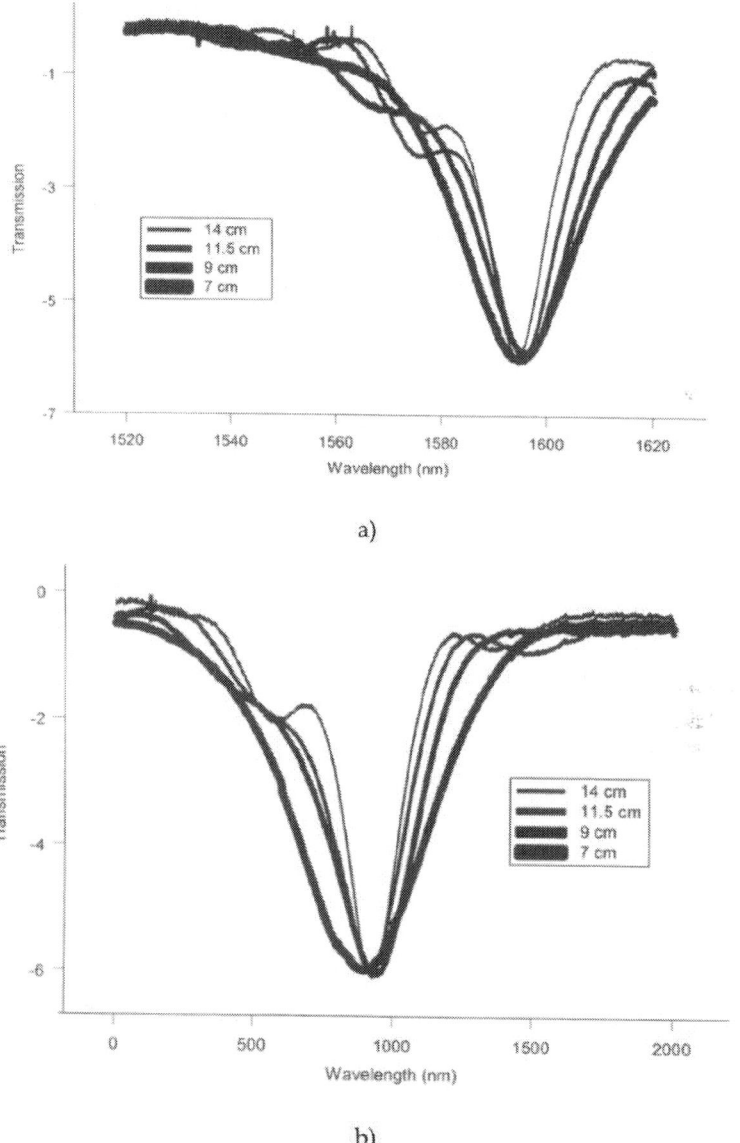

a)

b)

Figure 9: Measured bandwidth variation of filters at different peak wavelengths.

Double-Pass Configuration

One of the key problems in fiber-based AOTF is the low attenuation level of the notch filter. Superposing two or more filters according to Yun, Lee, Kim and Kim (1999) produced by multiple transducers can increase the attenuation level. But this method introduces a very high crosstalk in the device especially when the filter's peak: wavelengths are very near to each another and prove [18] highly impractical. Alternatively to improve the attenuation level of the filter, a double pass AO setup reported by Satorious, Dimmick, Burdge (2002) and Culverhouse, Yun, Richardson, Birks, Farwell, Russell (1997) can be used. In the new setup as in Fig 1.10, a 3-port circulator is added before and after the AO device. Light comes in from port 1 of circulator 1 and goes through the acoustic region and experiences mode conversion. The LP 11 coupled mode is converted back to the fundamental mode at the jacket of the fiber. The light rounds circulator 2 and goes through the AO device and experiences mode conversion again. The produced notch filter is observed using the OSA connected to the port 3 of circulator 1. Since the period of the acoustic inside the fiber is not changed, the light going through this region experiences mode conversion at the same wavelength of the incident and returning light.

The insertion loss (IL) of a double-pass is increased to less than 3 dB and the Polarization Dependent Loss (PDL) was less than 0.6 dB. IL was not intentionally increased to a high value here, because 2 FC/FC connectors were introduced in the setup to connect port 2 of both circulators to the AO device. Splicing the ports to the device will reduce the IL loss to values less than 1 dB. Using higher quality circulators can further reduce PDL of the notch filters. The filters however will be more expensive to fabricate.

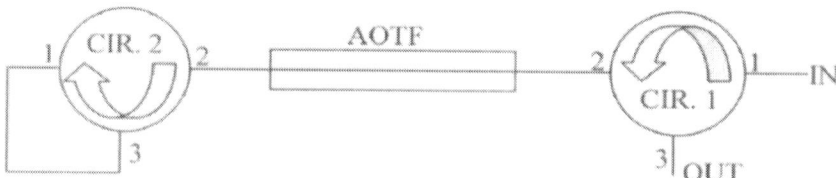

Figure 10: Double pass configuration to increase the maximum attenuation of the notch filter.

The AO band pass filter by Satorius *et. al.*, mentions side lobe suppression and maximum attenuation suppression using the double-pass configuration. Unlike the band pass filter, the notch filter will increase the side lobe level and maximum attenuation level using the double-pass configuration. The side lobe increment is not significant to exceed the bottom level of the main lobe of the notch filter.

From Fig. 11 the maximum attenuation of the notch filter was -28 dB for the double pass configuration and -12 dB for the single pass configuration. The maximum attenuation of the filter was increased to more than two times. The 3-dB bandwidth of the single pass AO device is 6.14 nm and the 3-dB bandwidth of the double pass AO device is 2.383 nm. This technique will be useful in producing narrow filters with high attenuation suitable in switching applications. However, there will be a frequency shift of 7 nm introduced using this setup.

The optical signal coupled from the slow mode (LP01) to the fast mode (LP11) will be downshifted in frequency when the acoustic wave is in the same direction as the optical signal. Frequency is shifted up when the fast mode is coupled to the slow mode for the same acoustic wave [19]. The frequency shift direction is reversed when the acoustic wave is in opposite direction with the optical signal as reported by Kim, Blake, Engan, and Shaw, (1986). In a double-pass setup, the optical signal is both the same and opposite direction to the acoustic wave, while in a single-pass setup, optical signal is maintained in the same direction as the acoustic wave. Thus, a frequency shift is observed in a double-pass setup.

GAIN FLATTENING FILTER

The technique to vary the 3-dB bandwidth of filter inside SMF is then extended as a dynamic gain equalizer for the gain profile of an Erbidium Doped Fiber Amplifier (EDFA). This is just one of the possible applications of AO interaction as efficient spectral shaping devices. Various efforts to dynamically control the gain flatness of the ASE spectrum using acousto-optic tuneable filters (AOTF) were well demonstrated. Passive gain equalization as reported by Vengsarkar, Pedrazzani, Judkins, Lemaire, Bergano, and Davidson (1996) is unable to encounter gain variations due to different input optical power of Wavelength Division Multiplexing (WDM) channels. Meanwhile, integrated AOTF as gain flattening filters have a serious limitation of high insertion loss and crosstalk problems.

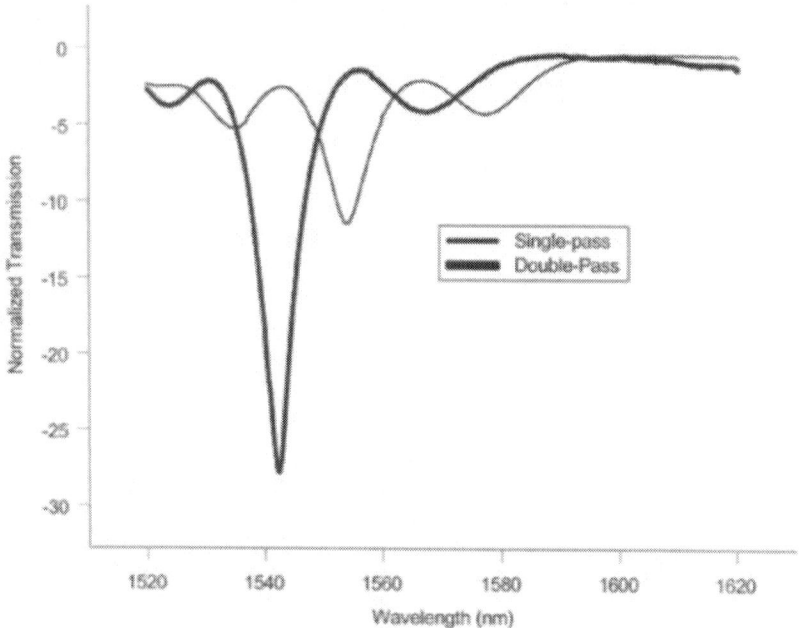

Figure 11: Measured normalized transmission spectrum using the double pass configuration. The result is compared with single pass configuration (refer to Fig4.1). There is a frequency shift of7 nm to the left using the double-pass setup.

Fiber-based AOTF by H.S. Kim, Park, and B.Y. Kim (1998) and the setup by Feced, Alegria, and Zervas (1999) uses two transducers with six synthesizers to obtain the desired spectral filters. In this technique, to shape the gain, the AOTF setup is using only one transducer and a single-taper. This is possible because the 3-dB [21-24] bandwidth of the filter we demonstrated can be varied on the same device. In our setup to flatten the gain profile of the Amplified Stimulated Emission (ASE) spectrum, an AOTF device with two frequency generators and a double-branched power combiner is used as inFig 12. The power combiner typically introduces a 3-dB loss to the system, thus higher Vp-p from the RF generator is needed to produce the filters for spectral shaping. Total insertion loss of the setup is less than 0.2 dB. For the measurement, the EDF A was used as the ASE source and the output spectrum measured [20] using an Optical Spectrum Analyser (OSA).

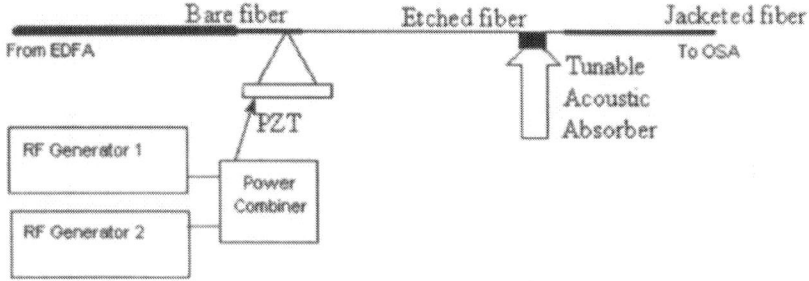

Figure 12: AOTF setup to flatten the gain of ASE spectrum.

The gain was flattened by changing the V_{p-p} level of the RF generator, and moving the tuneable acoustic absorber along the etched region of the SMF. The degree of freedom to shape the filter is very high, thus the necessity of cascading another AOTF to the setup is not needed. Fig. 13 shows the effect of shaping the filter on the Amplified Stimulated Emission spectrum of EDFA. Typically has it Amplified Stimulated Emission s peaks at 1532 run and 1550 run. For low gain, however there is a single broad peak at 1560 run. By using this method we show that, the [26-28] ASE spectrum can be flattened regardless of the peak's position and bandwidth using the same device. Since the tuning range is about 300 run, any Amplified Stimulated Emission spectrum that is lying from 1350 run to 1630 run can be successfully flattened using the same device.

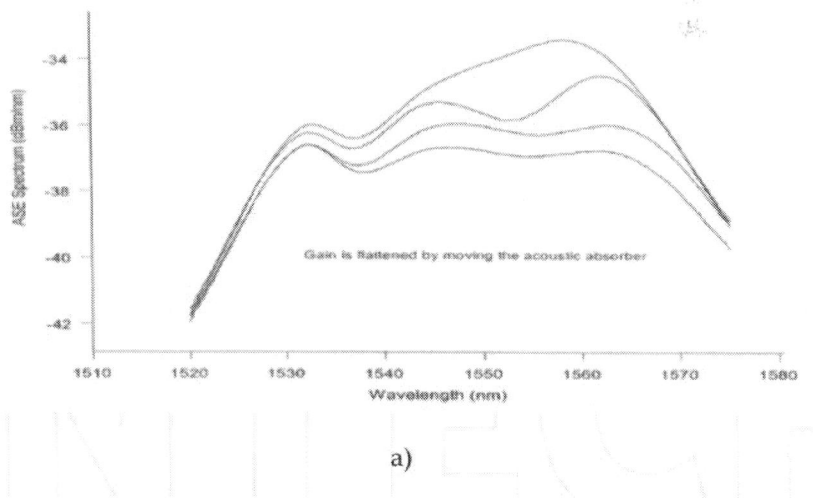

a)

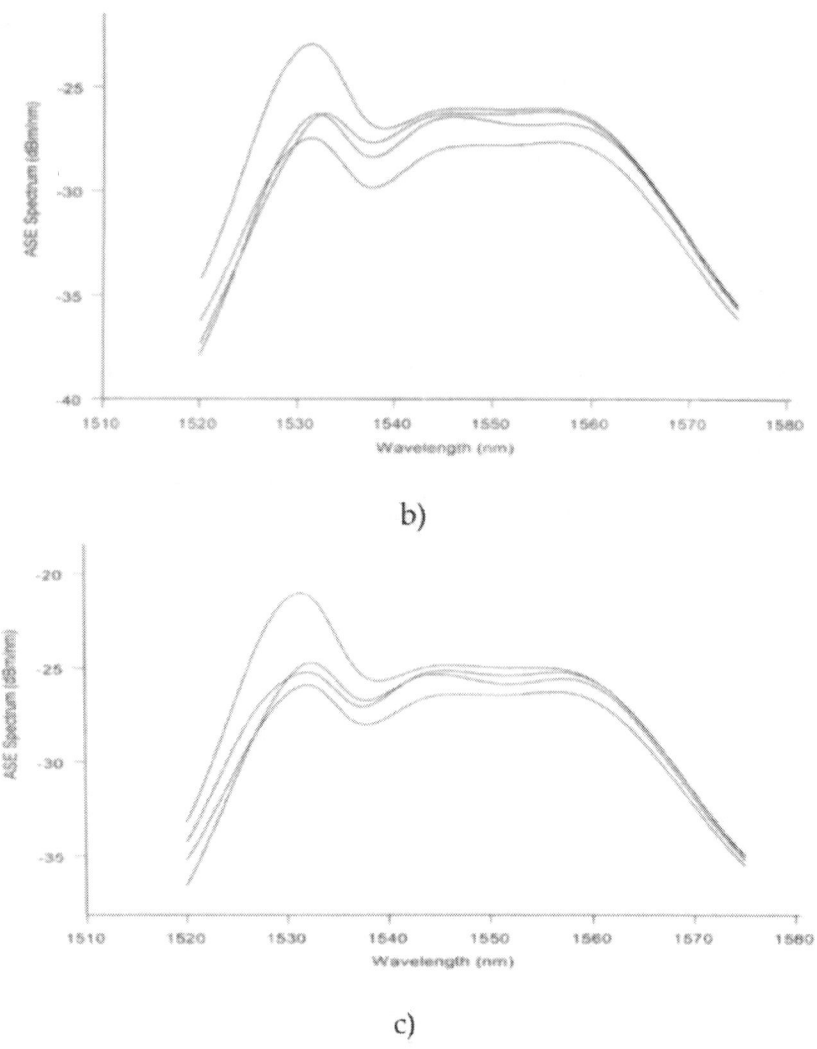

b)

c)

Figure 13: The effect of moving the tuneable acoustic absorber on the Amplified Stimulated Emission spectrum at various gain levels: a) low gain single peak at 1560 nm b) and c) high gain two peaks at 1532 nm and 1550 nm.\

Fig. 14 shows the flattened gain of ASE spectrum at various gain levels using this technique. For the lowest gain, at -30 dBm, which is achieved with a pump power of 96 mA, a broad filter is needed at 1545 nm; to obtain this; the tuneable acoustic absorber is positioned 14 cm after the AOTF device. The required resonant frequency to produce the coupling will be 990 kHz. A deeper notch is needed at 1532 nm; which is produced through the second frequency generator that is set at 993 kHz. The flattened gain is less than 0.8 dB. For gain at -25 dBm, which is achieved with a pump power of 150 mA, a filter is needed at 1556 nm; and a narrow deep notch is needed at 1532 nm; the required resonant frequency to produce the coupling respectively will be 986 kHz and 993 kHz. To obtain this, the tuneable acoustic absorber is positioned 16 cm after the AOTF device. The flattened gain is less than 0.9 dB.

Similarly for gain at -22 dBm, which is achieved with a pump power of 220 mA, a very deep filter is needed at 1532 nm and a small filter at 1545 nm. The resonant frequencies corresponding to these wavelengths are 990 kHz and 993 kHz respectively. To obtain the narrow filter, the tuneable acoustic absorber was set 17 cm after the AOTF device. And the measured flattened gain is less than 0.9 dB.Fig. 5 represents the notch filters obtained to flatten the gain of the Amplified Stimulated Emission (ASE) spectrum.

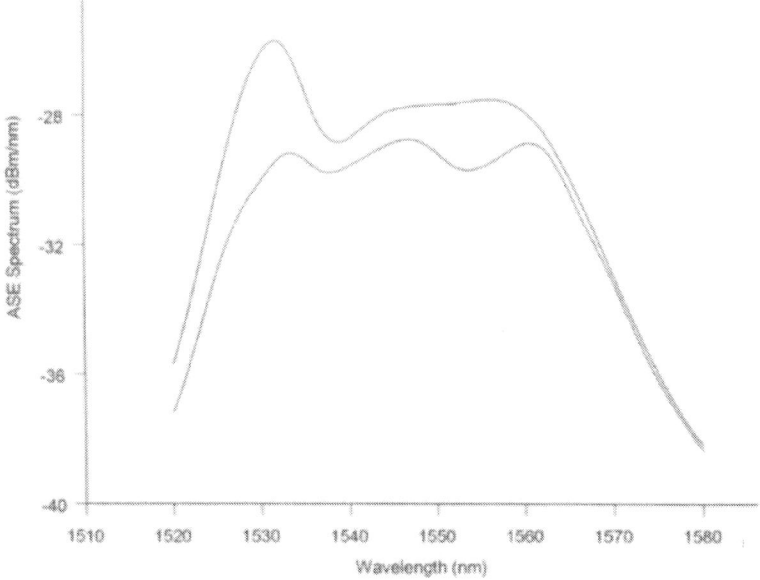

a)

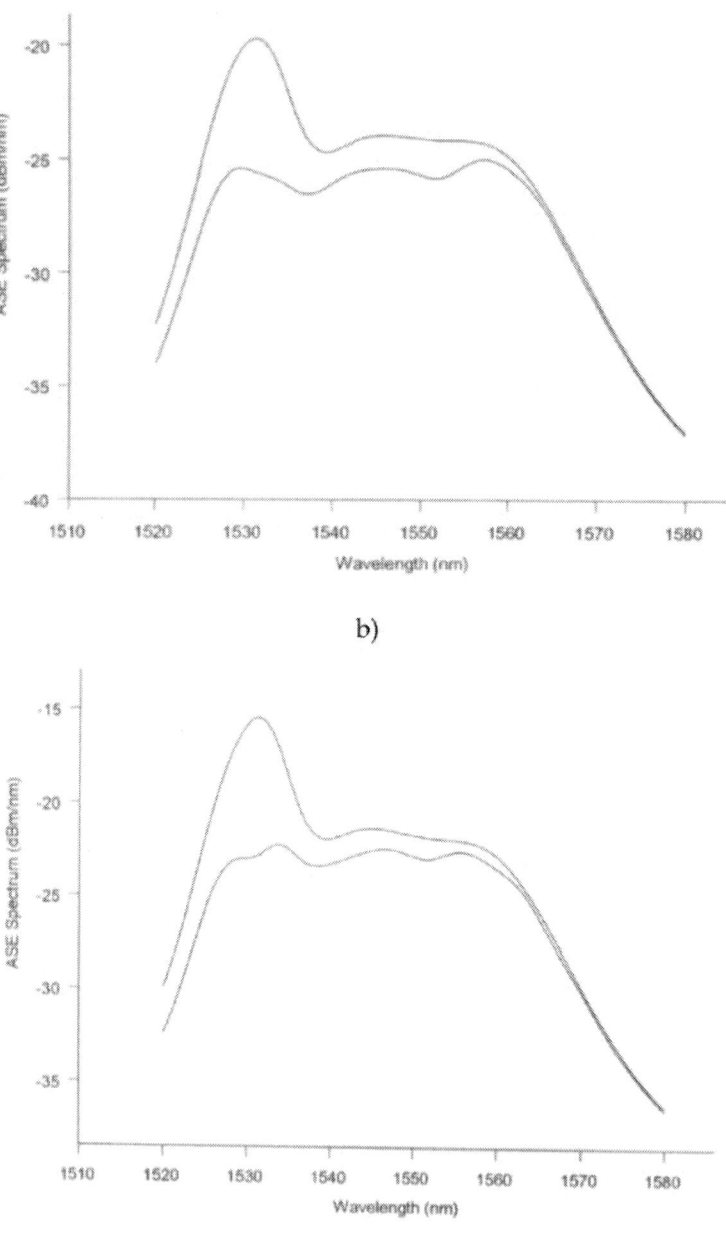

b)

c)

Figure 14: Gain profiles of the ASE spectrum and the flattened gain at various pump powers: a) 96 mA which has a gain of -30 dBm b) 150 mA which has a gain of -25 dBm and c) 220 mA which has a gain of -22 dBm.

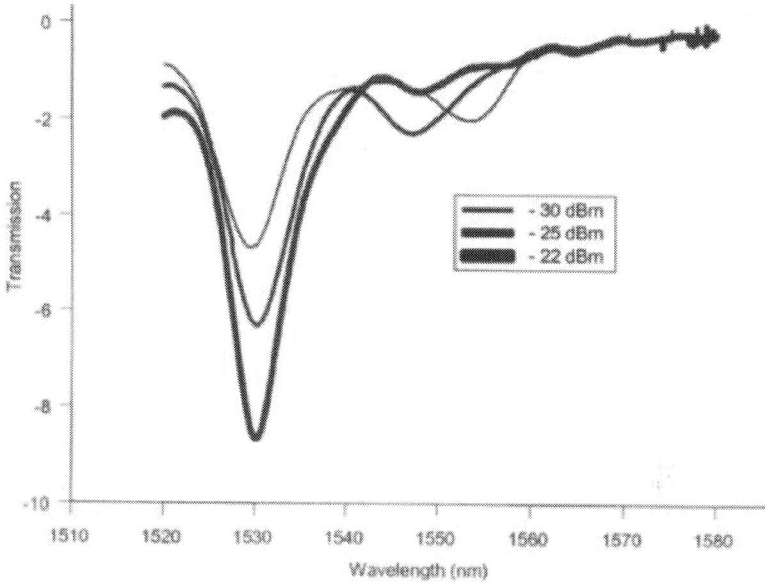

Figure 15: Corresponding filter spectrum to the flattened gain of various gain levels in Fig. 16.

CONCLUSION

The presence of acoustics inside the fiber will create a sequence of bends periodic in nature along the direction of its propagation. Core mode's energy is transferred to a cladding mode's, when it passes through the sequence of bends. The fiber jacket absorbs the coupled energy and this produces a notch filter observed using an optical spectrum analyzer.

Acoustic horn functions to transfer the acoustic wave of the transducer to the fiber. Aluminium horn is preferred over silica horn because it can be easily reproduced. Furthermore, its acoustic impedance almost matches that of silica's. To allow optimum transmission of acoustics to the fiber, the tip of the horn is made small, with its diameter matching that of silica's.

No resonance peaks were observed when the fiber is unetched First peaks are observed when the thickness of fiber is approximately 40 μ m. Overlap integral between the modes is not high in thicker fiber, meaning the transfer of acoustic wave to the fiber is not optimized. Thickness reductions in fibers are observed using a CCD camera. The characteristics of the resonance peaks can be controlled electrically using a RF generator. Voltage of the generator can be used to tune the attenuation depth of the resonance peaks. Frequency of the generator can be used to tune the peak wavelength. Frequency is inversely related to period of bends, thus higher frequencies will shift the peak to lower wavelengths. The 3-dB bandwidth of the resonance peaks can be adjusted by limiting the acoustic bend produced inside the fiber. Introducing a tunable acoustic absorber along the fiber can do this. Frequency used in all experiments was from 800 kHz to 1.1 MHz. All the coupled energy to produce the resonance peaks were to LP11 modes, mode conversion observed using a beam pro filer.

The power fed to cause resonance peaks can be reduced by reducing the thickness of the fiber to a value close to 20 μ m. Allowing light to pass through the acoustic bend region twice, as proposed in the double pass configuration, can increase the attenuation peaks. However, a frequency shift of 13 nm is observed because the light is passing through the bend in opposite directions.

As a spectral shaping tool, the attenuator is efficient as a gain flattening filter for an erbium doped amplifier. The peak of an amplified spontaneous emission at 1531 nm can be reduced to flat levels for various gains of the EDF A pump power. Insertion loss is less than 0.2 dB and polarization dependence loss is less than 0.4 dB.

REFERENCES

1. Sutharsanan VeeriahDesign and Characterisation of All Fiber Optical Filters", Master of Science Thesis, Faculty of Engineering, Multimedia University, Malaysia, Feb 2006
2. Abdulhalim 1Pannell C.N., (1993Acoustooptic in-fiber modulator using acoustic focusingIEEE Photonics Technology Letters999 EOF
3. A. A. Au, Q. Liu, C. H. Lin, H. P. Lee, 2004Effects of Acoustic Reflection on the Performance of a Cladding-Etched All-Fiber Acoustooptic Variable Optical AttenuatorIEEE Photonics Technology Letters150 EOF152 EOF
4. T. A. Birks, P. Russell, S. J. Culverhouse, D.O., (1992The Acousto-Optic Effect in Single-Mode Fiber Tapers and CouplersJournal of Lightwave Technology, 14 (11), 2519 EOF
5. D.O Culverhouse, ., S.H Yun, ., D.J Richardson, ., T.A Birks, , S.G Farwell, ., P Russell, . St, ., (1997). Low-loss all- fiber acousto-optic tunable filter. Optics Letters, 22 (2), 96-98

6. H. E. Engan, 2000Acousto-Optic Coupling In Optical FibersIEEE Ultrasonics Symposium, 625629

7. H. E. Engan, B. Y. Kim, J. N. Blake, H. J. Shaw, 1988Propagation and Optical Interaction of Guided Acoustic Waves in Two-Mode Optical FibersJournal of Lightwave Technology428 EOF436 EOF

8. H. E. Engan, D. Ostling, Kval Per 0., Askautrud Jan 0.. Wideband Operation of Horns for excitation of Acoustic Modes in Optical Fibers. 10th Optical Fibre Sensors Conference, 568571

9. R. Feced, c. Alegria, M. N. Zervas, 1999Acoustooptic Attenuation Filters Based on Tapered Optical FibersIEEE Journal of Selected Topics in Quantum Electronics1278 EOF1288 EOF

10. Y. Jung, S. B. Lee, J. W. Lee, K. Oh, 2005Bandwidth control in a hybrid fiber acoustooptic filter. Optics Letters, 30 (1), 84-86

11. G. Keiser, 1991Optical Fiber CommunicationsMcGraw-Hill, Inc., (2nd Edition) Kim RY, Blake J.N., Engan H.E., Shaw H.J., (1986). All-fiber acousto-optic frequency shifter. Optics Letters, 11 (6),389-391

12. H. S. Kim, S. H. Yun, H. K. Kim, N. Park, B. Y. Kim, 1998Actively Gain-Flattened Erbium-Doped Fiber Amplifier Over 35 nrn by Using All-Fiber Acoustooptic Tunable Filters. IEEE Photonics Technology Letters, 10 (6), 790-792

13. H. S. Kim, S. H. Yun, L. K. Kwang, B. Y. Kim, 1997All-fiber acousto-optic tunable notch filter with electronically controllable spectral profile.Optics Letters1476 EOF8 EOF

14. S. S. Lee, H. S. Kim, L. K. Hwang, S. H. Yun, 2003Highly-efficient broadband acoustic transducer for all-fibre acousto-optic devicesElectronics Letters

15. Q. Li, A. A. Au, C. H. Lin, E. R. Lyons, H. P. Lee, 2002An Efficient All-Fiber Variable Optical Attenuator via Acoustooptic Mode CouplingIEEE Photonics Technology Letters1563 EOF1565 EOF

16. Q. Li, X. Liu, H. P. Lee, 2002Demonstration of Narrow-Band Acoustooptic Tunable Filters on Dispersion-Enhanced Single Mode FibersIEEE Photonics Technology Letters1551 EOF1553 EOF

17. Q. Li, X. Liu, J. Peng, B. Zhou, E. R. Lyons, H. P. Lee, 2002Highly Efficient Acoustooptic Tunable Filter Based on Cladding Etched Single-Mode FiberIEEE Photonics Technology Letters337 EOF339 EOF

18. Q. Liu, K. S. Chiang, V. Rastogi, 2003Analysis of Corrugated Long-Period Gratings in Slab Waveguides and Their Polarization DependenceJournal of Lightwave Technology3399 EOF3405 EOF

19. J. D. Love, W. M. Henry, W. J. Stewart, R. J. Black, S. Lacroix, F. Gonthier, 1991Tapered Single-mode fibres and devices Part 1: Adiabaticity criteria.IEE Proceedings, 138, (5), 343 EOF

20. M. Monerie, 1982Propagation in Doubly Clad Single-Mode FibersIEEE Transactions on Microwave Theory and TechniquesMTT-30 (4), 381 EOF388 EOF

21. S. Mononobe, M. Ohtsu, 1996Fabrication of a Pencil-Shaped Fiber Probe for Near- Field Optics by Selective Chemical Etching. Journal of Lightwave Technology, 14 (10), 2231 EOF2235 EOF

22. D. Ostling, H. E. Engan, 1995Narrow-band acousto-optic tunable filtering in a twomode fiber. Optics Letters, 20 (11), 1247-1249

23. D. Ostling, H. E. Engan, 1995Spectral Flattening by an All-Fib er Acousto-Optic Tunable Filter. IEEE Ultrasonics Symposium, 837840

24. C. N. Pannell, B. F. Wacogne, Abdulhalim 1., (1995In-Fib er and Fiber Compatible Acoustooptic Components. Journal of Lightwave Technology, 13 (7),1429-1434

25. D. A. Satorius, T. E. Dimmick, G. L. Burdge, 2002Double-Pass Acoustooptic Tunable Bandpass Filter With Zero Frequency Shift and Reduced Polarization SensitivityIEEE Photonics Technology Letters1324 EOF1326 EOF

26. S. H. Yun, B. W. Lee, H. K. Kim, B. Y. Kim, 1999Dynamic Erbium-Doped Fiber Amplifier Based on Active Gain Flattening with Fiber Acoustooptic Tunable FiltersIEEE Photonics Technology Letters1229 EOF1231 EOF

27. S. H. Yun, B. Y. Kim, H. J. Jeong, B. Y. Kim, 1996Suppression of polarization dependence in a two-mode-fiber acousto-optic device.Optics Letters908 EOF10 EOF

28. S. H. Yun, H. S. Kim, 2004Resonance in Fiber-Based Acoustooptic Devices Via Acoustic Radiation to AirIEEE Photonics Technology Letters147 EOF149 EOF

CITATION

Abhilash Mandloi and Vivekanand Mishra (2013). Acoustics in Optical Fiber, Modeling and Measurement Methods for Acoustic Waves and for Acoustic Microdevices, Prof. Marco G. Beghi (Ed.), ISBN: 978-953-51-1189-4, InTech, DOI: 10.5772/54477.

Chapter 7

Cascading Multi-Hop Reservation and Transmission in Underwater Acoustic Sensor Networks

Jae-Won Lee and Ho-Shin Cho

School of Electronics Engineering, Kyungpook National University, Daegu 702-701, Korea;

ABSTRACT

The long propagation delay in an underwater acoustic channel makes designing an underwater media access control (MAC) protocol more challenging. In particular, handshaking-based MAC protocols widely used in terrestrial radio channels have been known to be inappropriate in underwater acoustic channels, because of the inordinately large latency involved in exchanging control packets. Furthermore, in the case of multi-hop relaying in a hop-by-hop handshaking manner, the end-to-end delay significantly increases. In this paper, we propose a new MAC protocol named cascading multi-hop reservation and transmission (CMRT). In CMRT, intermediate nodes between a source and a destination may start handshaking in advance for the next-hop relaying before handshaking for the previous node is completed. By this concurrent relaying, control packet exchange and data delivery cascade down to the destination. In addition, to improve channel utilization, CMRT adopts a packet-train method where multiple data packets are sent together by handshaking once. Thus, CMRT reduces the time taken for control packet exchange and accordingly increases the throughput. The performance of CMRT is evaluated and compared with that of two conventional MAC protocols (multiple-access collision avoidance for underwater (MACA-U) and MACA-U with packet trains (MACA-UPT)). The results show that CMRT outperforms other MAC protocols in terms of both throughput and end-to-end delay.

INTRODUCTION

Underwater acoustic sensor networks (UWSNs) have begun to draw the attention of researchers because of their potential use in a wide variety of applications, such as environmental monitoring, resource investigation, disaster prevention and recovery, navigation and military surveillance [1]. To implement these applications efficiently, it is important to understand the characteristics of an underwater channel and to design an efficient media access control (MAC) protocol that allows communication nodes to access the shared channel.

Unlike in terrestrial wireless communication, radio signals suffer severe path losses in the underwater environment; therefore, acoustic signals are typically employed in underwater communication. However, underwater acoustic links also suffer path losses, time-varying multi-path fading, motion-induced Doppler spread and aquatic noise [2]. Accordingly, when designing an underwater MAC protocol, new challenges that arise because of the unique characteristics of the underwater acoustic channel need to be carefully considered. In particular, the speed of sound under water is nearly 1500 m/s, which is five orders of magnitude lower than a radio signal's propagation speed of 3×10^8 m/s. The underwater acoustic channel is also characterized by a narrow and low bandwidth that results in low data rates. Consequently, most terrestrial MAC protocols for wireless sensor networks (WSNs) cannot be directly applied in the underwater environment, because they are designed for supporting high data rates with negligible propagation delay.

Nonetheless, there have been some studies that have tried to apply the existing terrestrial MAC protocols to underwater environments. Under conditions of light traffic load, a purely uncontrolled random access protocol, such as Aloha, has a lower packet delay, because it transmits directly whenever a packet is generated. However, because of the lack of a collision avoidance mechanism, Aloha generates a significant number of collisions as the traffic load increases. The throughput analysis of Aloha in the underwater environment was presented in [3] and [4]. To reduce the collisions of Aloha, Nitthita et al. proposed two Aloha-based protocols, namely, Aloha with collision avoidance (Aloha-CA) and Aloha with advance notification (Aloha-AN) [5]. These two protocols utilize the information obtained from the overheard packets to calculate the busy durations of neighboring nodes and avoid collisions accordingly. Unlike Aloha-based protocols, the carrier sense multiple access (CSMA) [6] makes a node listen to the channel before transmitting a packet, that is, a node may start transmitting if and only if it senses that the channel is idle. However, in a long propagation delay environment, the carrier sensing cannot indicate the real status of the channel, which means that the carrier sensing mechanism is not appropriate for the underwater environment.

Current research efforts on underwater MAC protocols strongly focus on the handshaking-based MAC protocols that reserve a channel by exchanging control packets, such as request-to-send (RTS) and clear-to-send (CTS). Existing handshaking-based underwater MAC protocols can be categorized into two types: sender-initiated and receiver-initiated. The multiple-access collision avoidance (MACA) [7] is a popular and representative sender-initiated MAC protocol that uses the three-way RTS/CTS/DATA handshake. In MACA, an exchange of RTS and CTS between sender and receiver takes place prior to data transmission. Hence, neighbors overhearing the control packets can defer their communication in order to avoid possible collisions that are addressed as a hidden-node problem. However, in the underwater environment, the simple exchange of RTS and CTS barely solves the hidden-node problem because of the long propagation delay of the acoustic channel.

To overcome this problem, Molins and Stojanovic proposed slotted floor acquisition multiple access (Slotted-FAMA) [8] that combines both carrier sensing and RTS/CTS handshake mechanisms. In this protocol, packets are transmitted at the beginning of a slot whose length is equal to the maximum propagation delay. Although the Slotted-FAMA can prevent collisions caused by hidden nodes, the excessive slot length decreases the throughput performance. Like the Slotted-FAMA, the distance-aware collision avoidance protocol (DACAP) proposed in [9] combines carrier sensing and RTS/CTS handshake mechanisms, but the nodes need not be synchronized. This enables a sender to use different handshake lengths for different receivers to minimize the average handshake duration. In addition, DACAP waits some time before transmitting the data packet to guarantee the absence of harmful collisions.

Another CSMA-based MAC protocol, named propagation delay aware protocol (PDAP), was proposed in [10]. PDAP aims at maximizing the bandwidth utilization by keeping track of neighboring transmissions to avoid collisions, thus enabling interleaved packet transmission between different pairs of users. In order to solve the problem of space-time uncertainty, a new class of MAC protocol, called Tone Lohi (T-Lohi), was proposed in [11]. T-Lohi uses short contention tones to reserve the channel for competing nodes. This tone-based reservation mechanism provides collision avoidance and low energy consumption. However, T-Lohi requires a node to be idle and listen to the channel for every contention round when competing for the channel, and because the listening period lasts for at least the maximum propagation delay time plus the time to detect the contention tone, it results in a low channel utilization [12].

Among MACA-based protocols, MACA for underwater (MACA-U) [13] is the basic and reference protocol that revises the state transition rules that account for the long propagation delay. In [14], Liao and Huang proposed the spatially fair MAC (SF-MAC) protocol that concerns not only the collisions, but also the unfairness problem caused by the long propagation delay. SF-MAC prevents collisions by postponing the transmission of the CTS packet. The receiver collects RTS packets from all the potential senders during the RTS contention period and determines the earliest transmitter, achieving a higher degree of fairness. However, SF-MAC has a long, fixed RTS contention period, which critically affects channel utilization. To improve the channel utilization, Guo *et al.* proposed the adaptive propagation-delay-tolerant collision-avoidance protocol (APCAP) [15] that enables a sender to perform other functions during the large time gap between the transmission of RTS and the corresponding CTS reception, which is called MAC level pipelining. However, APCAP requires a time synchronization and a complicated process for MAC level pipelining.

A delay-aware opportunistic transmission scheduling (DOTS) protocol [16] uses passively obtained local information (neighboring nodes' propagation delay map) to increase the chances of concurrent transmissions while reducing the likelihood of collisions. Another way to improve the channel utilization is a packet-train approach. Chirdchoo *et al.* proposed a MACA-based MAC protocol with packet-train to multiple neighbors (MACA-MN) [17]. MACA-MN improves channel utilization by sending multiple packets to multiple neighbors in each round of handshake. MACA-U with packet trains (MACA-UPT) was also introduced in [18]. MACA-UPT is derived from MACA-U, except that a sender transmits multiple data packets in a single handshake in the former. Recently, Hai-Heng Ng *et al.* proposed a bidirectional concurrent MAC (BiC-MAC) protocol [18], wherein a sender-receiver pair simultaneously transmits data packets to each other, which improves the channel utilization. Hai-Heng Ng *et al.* also proposed a MAC protocol using reverse opportunistic packet appending (ROPA) [19], which is a hybrid of sender-initiated and receiver-initiated MAC protocols. ROPA improves channel utilization by enabling a sender to coordinate multiple neighbors to opportunistically transmit (append) their data packets. After the sender finishes transmitting its data packets, it starts to receive incoming appended data packets. However, in ROPA, more control packet exchange is needed; therefore, more collisions may occur.

On the other hand, in [20], Chirdchoo *et al.* proposed the receiver-initiated packet train (RIPT) protocol that falls into the category of the receiver-initiated MAC protocols. When a node wishes to become a receiver, it initiates the four-way ready-to-receive (RTR)/SIZE/ORDER/DATA handshake that schedules the packets from multiple neighbors to arrive at the receiver in a packet train. Although RIPT can get multiple data packets from neighbors, the four-way handshake takes a long time to receive the first packet train at the receiver node, especially in the underwater environment.

As described above, the long propagation delay, which is a major feature to be considered in the case of underwater acoustic channels, makes it difficult to design underwater MAC protocols. In particular, in handshaking-based MAC protocols, the exchange of control packets is time-consuming, resulting in a large signaling overhead. Furthermore, in the case of multi-hop relaying in a hop-by-hop handshaking manner, the end-to-end delay is significantly increased. Therefore, this paper proposes a new underwater MAC protocol, named cascading multi-hop reservation and transmission (CMRT), to address the abovementioned problems. The CMRT protocol reserves the multi-hop channels at once by cascading reservation control packets and delivers the data packets in the same way until they reach the destination without stopping at intermediate nodes. This multi-hop reservation approach is different from what conventional MAC protocols employ for multi-hop transmission as explained above. In addition, CMRT adopts a packet-train method [17] to improve channel utilization by sending multiple data packets together with only one handshaking signal. In this way, CMRT is able to reduce the control packet exchange time and accordingly increase the throughput compared with conventional MAC protocols. The main contributions of this paper can be summarized as follows:

1. Propose a cascading multi-hop reservation-based MAC protocol for UWSNs with a long propagation delay to significantly reduce the end-to-end delay and improve channel utilization.
2. Compare the performance with conventional MAC protocols in terms of throughput and end-to-end delay.
3. Propose a new RTS attempt triggering method that adaptively changes the batch size of data packets transmitted with a single reservation.

The rest of the paper is organized as follows. Section 2 presents the problem statements. In Section 3, we explain the proposed protocol design, including a new RTS attempt strategy. We present simulation results and their discussions in detail in Section 4. Finally, the conclusions are provided in Section 5.

PROBLEM STATEMENTS

The long propagation delay of the underwater acoustic channel poses challenges for the design of MAC protocols, such as space-time uncertainty and the hidden-node problem. Furthermore, the end-to-end delay is substantially increased in multi-hop relaying. In this section, we describe these problems in detail.

Space-Time Uncertainty

Figure 1a illustrates a collision that occurs in RF-based terrestrial WSNs where the propagation delay is negligible and the y-axis denotes the distance between nodes. When Nodes A and C are transmitting packets at the same time, the packets collide at destination Node B. Such collisions can be avoided by scheduling in such a way that the durations of the transmission time do not overlap. That is, we have to consider only the transmission time uncertainty.

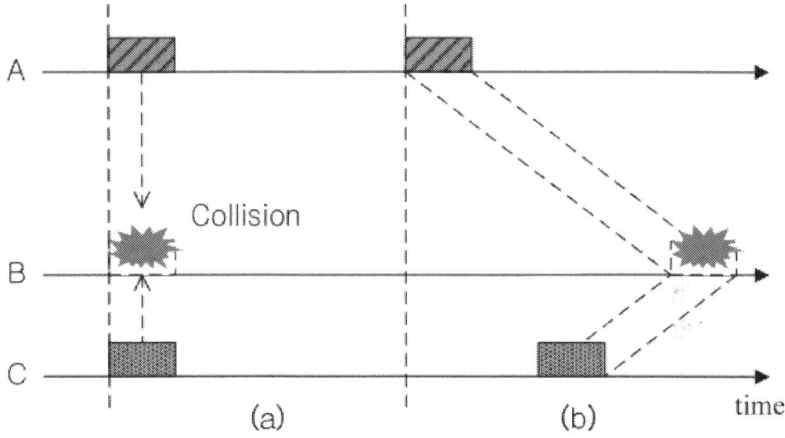

Figure 1: Space-time uncertainty: (a) terrestrial RF channel; and (b) underwater acoustic channel.

On the other hand, in the case of UWSNs, the long propagation delay of the acoustic signal makes it more complicated to avoid any collisions, because we have to consider not only the transmission time, but also the distance (space) between nodes. Figure 1b shows an example where two packets transmitted from Nodes A and C at different times collide at Node B. We call such a two-dimensional uncertainty in determining a collision at the receiver as space-time uncertainty [4].

Hidden-Node Problem in UWSNs

In conventional handshaking protocols for collision avoidance (CA), the source node makes a channel reservation by sending an RTS control packet. The destination node replies to the RTS with a CTS control packet that can be overheard by neighbors (potential interferers), so that they recognize that the channel will be reserved during a certain amount of time. Accordingly, the source node can transmit data packets to the destination node without collisions. This is the basic method adopted in CA protocols for preventing possible collisions caused by hidden nodes.

However, the long propagation delay in the underwater acoustic channel introduces a new kind of hidden-node problem, as shown in Figure 2.

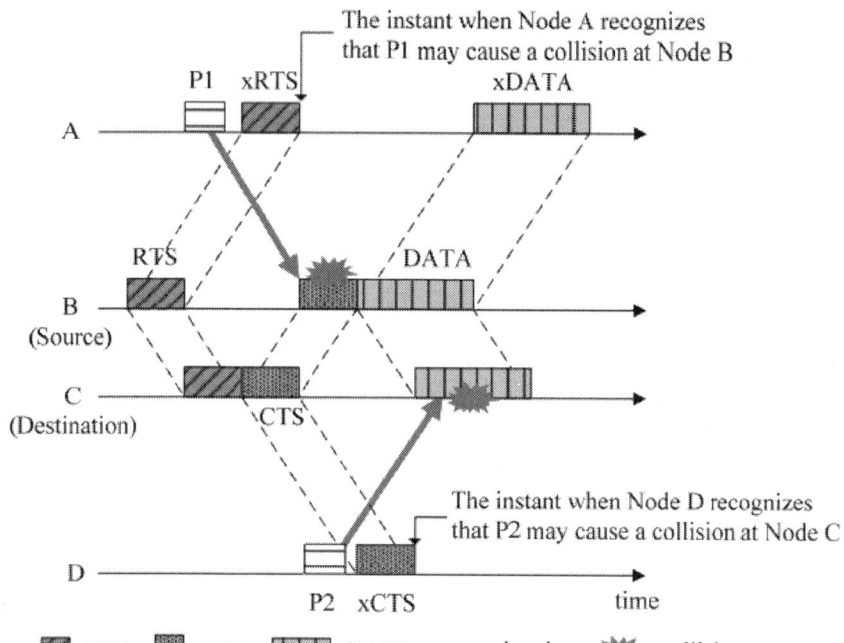

: RTS, : CTS, : DATA, x: overhearing, : collision.

Figure 2: Hidden-node problem in the underwater acoustic channel. RTS, request-to-send; CTS, clear-to-send.

In the underwater acoustic channel, some nodes may detect the channel reservation after transmitting control packets (e.g., P1 and P2 in Figure 2). This may cause possible collisions at source and destination nodes as denoted by the solid arrows in Figure 2. We call this unexpected collision caused by the long propagation delay the hidden-node problem in the underwater acoustic channel. In Figure 2, Nodes A and D become hidden nodes.

PROPOSED CMRT PROTOCOL

System Description

We consider a multi-hop network where all nodes are equipped with half-duplex and omni-directional acoustic modems. It is assumed that every node knows the inter-nodal distance to its neighbors within a one-hop range and keeps a list of those with which it can establish a bi-directional link. During the network initialization phase, the inter-nodal distance is obtained by using round-trip time (RTT) measurements of control packets or by sharing some information among neighbors [21]. It is also assumed that every node has the routing table to facilitate multi-hop relay.

Definition of States

In CMRT, a node shifts between six different states, namely, Idle, Wait_Resp (Wait for RESPonse), Delay_Data (Delay Data transmission), Wait_Data (Wait for Data reception), Data_Rx (Data Reception) and Silence.

Figure 3 illustrates the individual states that may occur in the CMRT procedure.

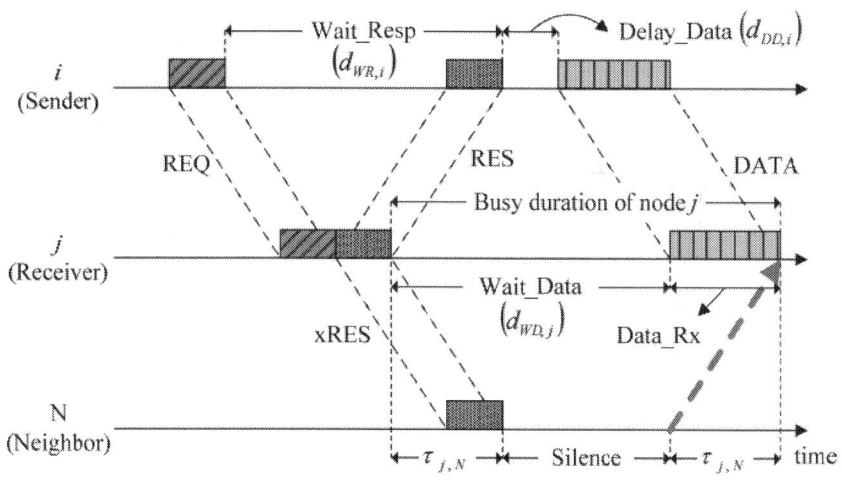

REQ ▨ : Request control packet, RES ▦ : Response control packet

Figure 3: The six different states of a node.

1. Wait_Resp is a state where a sender waits for a response to a request control packet (e.g., RTS) from a receiver. The sender stays in the Wait_Resp state directly after transmitting a request control packet until receiving a response control packet (e.g., CTS). If the sender does not receive a response control packet within the duration of Wait_Resp state, it will transit to the Idle state.

2. Delay_Data is a state where a sender delays data transmission to avoid possible collisions caused by the hidden nodes. After receiving a response control packet from the receiver, the sender enters the Delay_Data state and remains there until it starts transmitting data packets. The length of the Delay_Data state should be elaborately calculated, and the calculation procedure will be presented in Section 3.2.

3. Wait_Data is a state where a receiver waits for data packets from a sender. The receiver enters the Wait_Data state directly after transmitting a response control packet and remains there until it starts receiving data-packets.

4. Data_Rx is a state where a receiver receives data packets.

5. Silence is a state where neighbors who overheard the exchange of control packets for channel reservation remain silent, doing nothing so that they do not cause collisions. Neighbors enter the Silence state after overhearing the control packets involved in other nodes' channel reservation until the channel becomes free of reservation. The Silence state ensures that any transmissions from neighbors arrive after data reception is completed at a receiver, as denoted by the dotted arrow inFigure 3.

6. The Idle state includes the remaining cases not belonging to the five states described above.

The length of each state that indicates the time duration of node i staying in the corresponding state is listed in Table 1.

Table 1: Description of the length of each state

Notation	Description
$d_{WR,i}$	Length of the Wait_Resp for node i
$d_{DD,i}$	Length of the Delay_Data for node i
$d_{WD,i}$	Length of the Wait_Data for node i

Length of the Silence State

First, we define the busy duration as an interval between when a control packet is sent to neighbors and when a responding control packet (e.g., CTS) or a data packet is received as a reply from the neighbors. In the case of a sender (node i in Figure 3), the reply is carried out by a responding control packet. Thus, the busy duration of node i is given by the interval between the end of REQ (request) transmission and the end of RES (respond) reception as:

$$d_{busy,i} = d_{WR,i}$$

(1)

On the other hand, in the case of a receiver or a relay (node j in Figure 3), the reply is carried by a data packet. Thus, as shown in the Figure 3, the busy duration of node j includes not only the Wait_Data, but also the data reception time denoted by d_{DATA} as:

$$d_{busy,j} = d_{WD,j} + d_{DATA}$$

(2)

Every node specifies its busy duration inside the control packets. For example, in Figure 3, REQ and RES contain the busy durations for node i and j, respectively. Overhearing RES from node j, its neighbor, Node N, can easily calculate the length of Silence from:

$$d_{silence} = d_{busy,j} - 2\tau_{j,N}$$

(3)

where $\tau_{i,j}$ is the propagation delay between nodes i and j. In the same way, all neighbors receive information regarding how long they have to stay in Silence. Whenever a node in Silence overhears another neighbor's control packet, it extends its Silence duration accordingly.

Channel Occupancy Priority

In general, relay nodes handle two types of data packets: those generated by themselves, called domestic data packets, and those relayed from the neighbors, called foreign data packets. It is assumed that a foreign data packet has priority to occupy the channel over a domestic data packet. Such a policy is named the foreign-first policy. Each node manages two separate buffers, one each for domestic and foreign data packets. Let $N^{i \to j}$ be the number of data packets destined for node j and stored in the buffer of node i. Now,

$$N^{i \to j} = N^{i \to j}_{dom} + N^{i \to j}_{frg}$$

(4)

where $N^{i \to j}_{dom}$ and $N^{i \to j}_{frg}$ are the numbers of domestic and foreign data packets destined for node j and stored in the buffer of node i, respectively. $N^{i \to j}$ has a limited capacity of N_{max}. In terms of priority among foreign data packets, those belonging to the dominant set that contains a group of data packets destined to node k, such that $\underset{k}{\arg\max}\left[N^{i \to k}_{frg}\right]$, are transmitted first. This priority policy is named the dominant-first. Furthermore, inside the dominant set, data packets that have traveled by the largest number of hops until that instant are given the highest priority to occupy the channel. This priority policy is named the oldest-first.

Cascading Multi-Hop Reservation

Figure 4 shows a scenario of CMRT operation. It is assumed that two relays R1 and R2 exist between source S and destination D. A multi-hop relay begins with the source S staying in the Idle state by transmitting RTS to relay R1. After transmitting RTS, Node S enters the Wait_Resp state. The RTS packet contains the following information: (1) the address of the final destination (FD), Node D in this example; (2) batch size, the number of data packets to be transmitted, B_{size}; (3) the busy duration of Node S, $d_{busy,S}$; and (4) hop count to denote the number of hops from the source node, k_S. The value of the hop count will be increased by one as the channel reservation progresses.

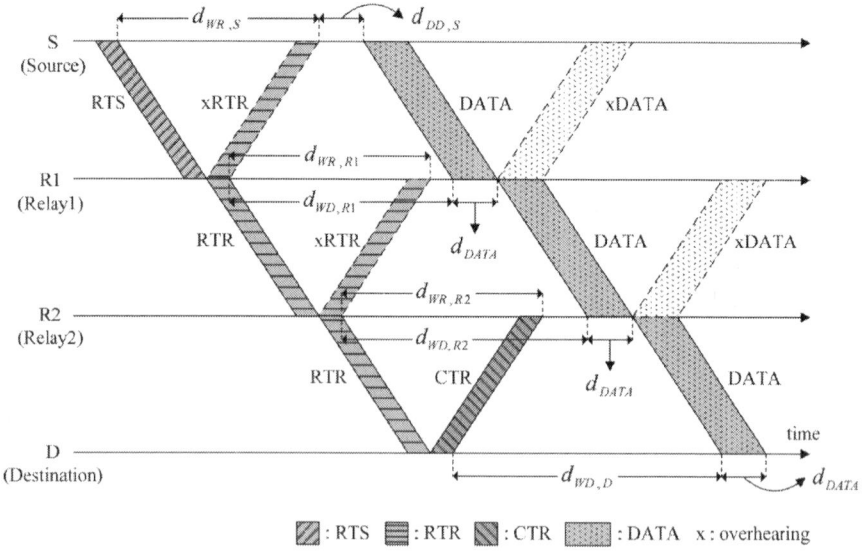

Figure 4: Operation of the cascading multi-hop reservation and transmission (CMRT) protocol.

Upon receiving RTS, the relay node R1 transmits a control packet named request-to-reserve (RTR) to the next node in order to reserve the channel for the next hop. Here, RTR is a newly introduced control packet in CMRT and is paired with a responding control packet named clear-to-reserve

(CTR) similar to the pairing of RTS with CTS. The RTR packet also contains the same information as RTS, [FD, B_{size}, $d_{busy,R1}$, k_{R1}], where $k_{R1} = k_S + 1$. The RTR is used not only to reserve the channel for the next hop, but also to respond to RTS/RTR of the previous hop to allow backward overhearing. In Figure 4, when Node R1 relays RTR to Node R2 in the forward direction, Node S overhears RTR in the backward direction, which is denoted by xRTR, to recognize that the node's previous request (RTS) was successfully sent and processed for the next-hop relay. After relaying RTR, Node R1 enters Wait_Resp and Wait_Data states at the same time. Accordingly, unlike a sender (Node S), Node R1 would not transit to the Idle state immediately, even if it does not receive a response control packet (xRTR) from node R2 within the duration of Wait_Resp state. Instead, Node R1 will stop the data forwarding and transit to the Idle state after the Data_Rx state regardless of whether it successfully receives a train of data packets. All relay nodes work in the same manner as Node R1. Destination D stops relaying RTR and instead transmits CTR to the previous relay node as a response to RTR. The CTR packet (as well as the CTS packet) contains the information about the busy duration of destination D ($d_{busy,D}$). Note that the RTR plays a key role here for cascading reservation information through multiple hops, thus efficiently reducing handshaking and data delivery times.

Source S delays data transmission for the length of Delay_Data ($d_{DD,S}$) to avoid causing possible collisions with the hidden nodes. For the simple case of two-hop relaying (source → relay, relay → destination) illustrated in Figure 5, the following relation between timing parameters is obtained:

$$2\tau_{S,R} + d_{DD,S} = 2\tau_{R,D} + T_{control}$$

(5)

where $T_{control}$ is the common transmission time of all control packets. Let τ_{max} be the maximum propagation delay between nodes that corresponds to the transmission range. The length of Delay_Data for the worst case ($\tau_{R,D} = \tau_{max}$) is now obtained as:

$$d_{DD,S} = 2(\tau_{max} - \tau_{S,R}) + T_{control}$$

(6)

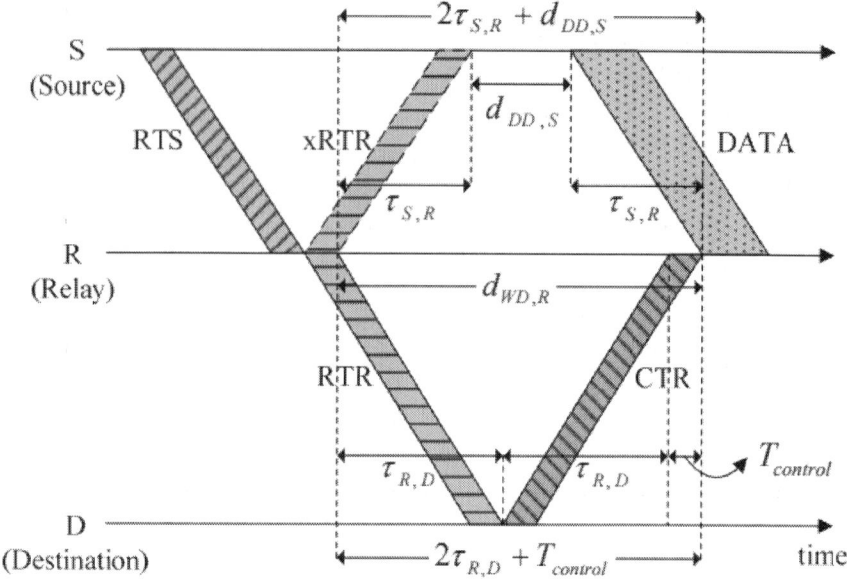

Figure 5: Determination of $d_{DD,S}$.

Data Transmission Using the Packet-Train Method

To increase channel utilization, CMRT adopts a packet-train method [17] where multiple data packets are sent in a row by handshaking once. Figure 6 illustrates how the packet-train method is used in CMRT for the case of B_{size} = 3. Source S sequentially transmits a train of data packets to the next relay node without any interval between packets. Similarly, relay Nodes R1 and R2 also forward the train without delay because the multi-hop channels to the destination are already reserved. If a relay node does not receive a train of data packets within the duration of the Data_Rx state, it would transit to the Idle state after the Data_Rx state ends.

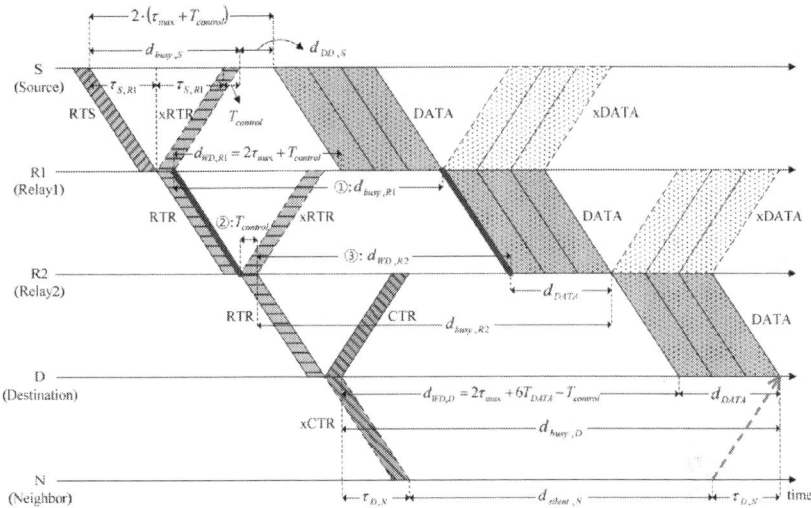

Figure 6: Data transmission using a packet train.

Depending on the batch size, the busy duration for the different types of nodes, namely, source, relay and destination, should be determined. On the basis of Equation (1), the busy duration of Node S inFigure 6 is given by:

$$d_{busy,S} = 2\tau_{S,R1} + T_{control} \tag{7}$$

On the basis of Equation (2), the busy duration of R1 is:

$$d_{busy,R1} = d_{WD,R1} + d_{DATA} \tag{8}$$

where $d_{WD,R1}$ is obtained from Figure 5 and Equation (6) by:

$$
\begin{aligned}
d_{WD,R1} &= 2\tau_{S,R1} + d_{DD,S} \\
&= 2\tau_{S,R1} + 2\left(\tau_{max} - \tau_{S,R1}\right) + T_{control} \\
&= 2\tau_{max} + T_{control}
\end{aligned} \tag{9}
$$

and:

$$d_{DATA} = B_{size} \cdot T_{DATA} \tag{10}$$

where T_{DATA} is the transmission time of a single data packet. In the same way as Node R1, the busy duration of R2 is:

$$d_{busy,R2} = d_{WD,R2} + d_{DATA} \tag{11}$$

where the length of Wait_Data of R2, $d_{WD,R2}$ (denoted by ③ in Figure 6), is directly obtained from $d_{busy,R1}$ (denoted by ①, the time duration between the two bold lines in Figure 6) and $T_{control}$ (denoted by ②). That is,

$$d_{WD,R2} = d_{busy,R1} - T_{control} \quad (③ = ① - ②)$$

(12)

Substituting Equations (10) and (12) for Equation (11), the busy duration of R2 is finally given by:

$$
\begin{aligned}
d_{busy,Ri} &= d_{WD,Ri} + d_{DATA} \\
&= \left(d_{busy,Ri-1} - T_{control}\right) + B_{size} \cdot T_{DATA} \\
&= 2\tau_{max} + i \cdot B_{size} \cdot T_{DATA} - (i-2) \cdot T_{control}
\end{aligned}
$$

(13)

In the same way, we can generalize the busy duration of relay node Ri as:

$$
\begin{aligned}
d_{busy,Ri} &= d_{WD,Ri} + d_{DATA} \\
&= \left(d_{busy,Ri-1} - T_{control}\right) + B_{size} \cdot T_{DATA} \\
&= 2\tau_{max} + i \cdot B_{size} \cdot T_{DATA} - (i-2) \cdot T_{control}
\end{aligned}
$$

(14)

where the value of i can be determined from the hop-count included in RTS/RTR packets. Similarly, the busy duration of the destination when k relay nodes exist between the source and destination nodes is given by:

$$d_{busy,D} = 2\tau_{max} + (k+1) \cdot B_{size} \cdot T_{DATA} - (k-1) \cdot T_{control}$$

(15)

Handshaking Triggering and Back-Off Algorithm

In conventional CA protocols, each node starts the handshaking procedure under two types of conditions: (1) when the buffer becomes full (batch-by-size); and (2) when a predefined timer is expired (batch-by-time). In [18], a hybrid scheme combining "batch-by-size" and "batch-by-time" conditions was used. In the batch-by-size scheme, the proper value of the batch size (B_{size}) depends on traffic conditions. If B_{size} is too large under light traffic, the node spends too much time in waiting until the buffer becomes full. On the other hand, if B_{size} is too small under heavy traffic, the node tries handshaking too frequently, resulting in a heavy signaling load. In CMRT, we propose a new scheme named "batch-by-adaptive-size" where B_{size} of a given node I is adaptively changed according to the traffic condition as:

$$B_{size} = N^{i \to j}$$
(16)

where:

$$j = \arg\max_{k} \left(N^{i \to k} \right)$$
(17)

A batch-by-adaptive-size scheme is capable of working adaptively under varying traffic load conditions.

For RTS trials, CMRT adopts the binary exponential back-off (BEB) algorithm specified in the IEEE 802.11 standard [22]. In the BEB algorithm, a sender doubles its back-off counter (B_{cnt}) with the upper bounds of B_{max} when an RTS fails. On the other hand, upon successful transmission, a node resets its back-off counter to the minimum value of B_{min}. The duration of the back-off is selected randomly in the range of zero to the back-off interval ($B_{interval}$), which can be expressed as:

$$B_{interval} = random(0, B_{cnt}) \times \tau_{max}$$
(18)

SIMULATIONS AND RESULTS

Simulation Model

An event-driven network simulator was developed using MATLAB. As shown in Figure 7, a multi-hop topology is considered to have 36 static nodes placed in a 5000 × 5000 m^2 square area with a grid spacing of 1000 m. All of the nodes are assumed to have the same transmission power and, accordingly, the same transmission range (1.5-times the grid spacing), such that each node has exactly eight neighbors within its range (the dotted circle in Figure 7). Each node generates data packets according to the Poisson process with an arrival rate λ_{node} (packets/s) and randomly selects a destination with equal probability. For multi-hop transmission, we apply the static routing where each node uses a manually configured routing entry. The acoustic channel is assumed to be error-free; that is, packet losses occur only in the case of packet collisions. The system parameters for simulation are summarized in Table 2. The transmission rate is referenced from the LinkQuest medium range acoustic mode [23].

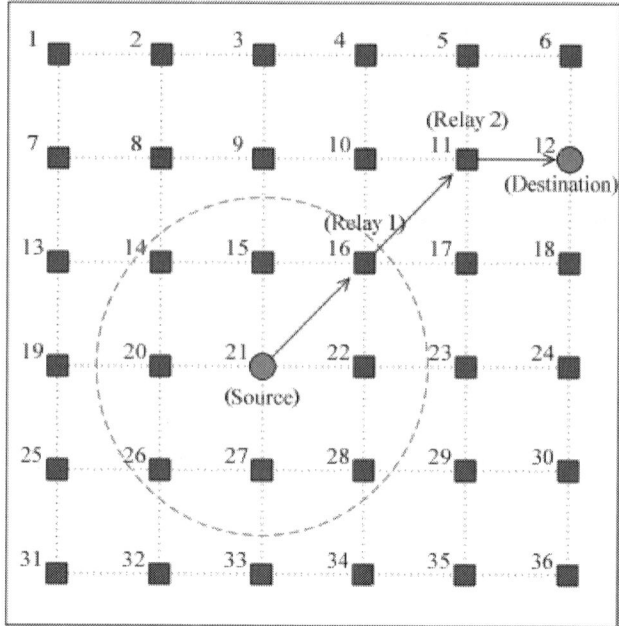

Figure 7: The network topology for simulations.

Table 2: System parameters

Parameter	Value
Transmission rate	9600 bps
Size of data packet	1200 bits
Size of control packet	120 bits
Minimum back-off counter (B_{min})	1
Maximum back-off counter (B_{max})	64
Capacity of buffer (N_{max})	300 packets
Acoustic propagation speed	1500 m/s

Simulation Results

The CMRT is compared with the two conventional MAC protocols, MACA-U [13] and MACA-UPT [18], and the single-hop repeated version of CMRT (CMRT-S) in terms of the normalized throughput per node and end-to-end packet delay. CMRT-S is a modified version of CMRT, where a single-hop transmission of CMRT is repeated multiple times in a hop-by-hop manner until the destination, as shown in Figure 8. Unlike the hop-by-hop application of previous protocols, such as MACA-U and MACA-UPT, CMRT-S uses the collision-escape mechanism that originated from CMRT, according to which after receiving a CTS, the source waits for a certain amount of time before transmitting data packets.

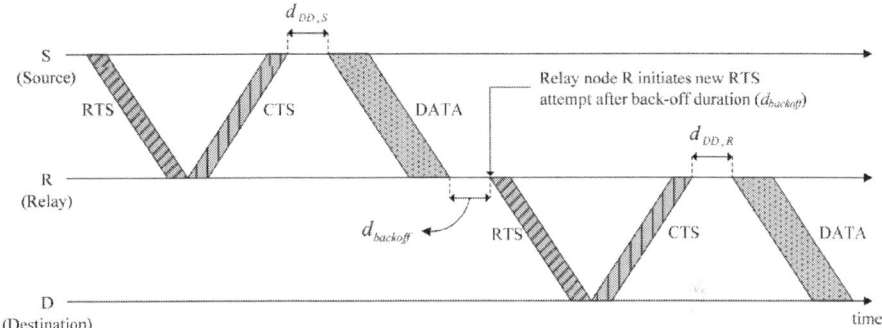

Figure 8: Operation of CMRT-S (single hop).

The normalized throughput per node is defined as:

$$\gamma = \frac{1}{N} \cdot \frac{\sum_{i=1}^{N} r_i \cdot B_{Data}}{t_{sim}}$$

(19)

where B_{Data} is the size of a data packet in bits and N is the total number of nodes in the network. Depending on the position along the data-packet relay route, the node could be a source, a relay or a destination. r_i is the number of data packets successfully received by destination node i, and t_{sim} is the simulation time. As another performance measure, the end-to-end packet delay is defined as the time duration from when a data packet is generated at a source to when it is successfully received at a destination. Let Ω be the set of data packets that arrive successfully at the destination. The size of Ω is given by $N(\Omega) = \sum_{i=1}^{N} r_i$, and each element of Ω has a different end-to-end packet delay, $t_{delay,j}$, $j = 1,2, \dots ,N(\Omega)$. Thus, the average end-to-end packet delay is defined as:

$$\overline{t_{delay}} = \frac{\sum_{j=1}^{N(\Omega)} t_{delay,j}}{N(\Omega)}$$

(20)

Regarding the channel occupancy priority, the three policies of foreign-first, dominant-first, and oldest-first described in Section 3.1.3 are applied to all protocols for comparison.

Comparison of CMRT with Other MAC Protocols

Figure 9a shows the normalized throughput per node (hereafter referred to as throughput) for various offered loads per node (λ_{node}, hereafter referred to as offered load), and Figure 9b shows the average end-to-end packet delay (hereafter referred to as delay) performance. In most cases of the offered load, CMRT exhibits the best performance in terms of both throughput and delay. That is because CMRT is able to significantly reduce the time spent in handshaking and data transmission by means of cascading channel-reservation and data-transmission over multiple hops and the packet-train method. Additionally, CMRT considers the hidden-node problem when scheduling the transmission time. Unlike CMRT, the three other protocols (CMRT-S, MACA-UPT and MACA-U) conduct multi-hop transmission by adopting the conventional way of hop-by-hop relaying, in which the next-hop relay starts only after completion of the previous-hop relay. The reason CMRT-S exhibits better performance than MACA-UPT and MACA-U is that CMRT-S is capable of handling the hidden-node problem by postponing the data transmission after receiving a CTS. MACA-U, which does not use the packet-train method, exhibits the worst performance. The features of the schemes aforementioned are summarized in Table 3.

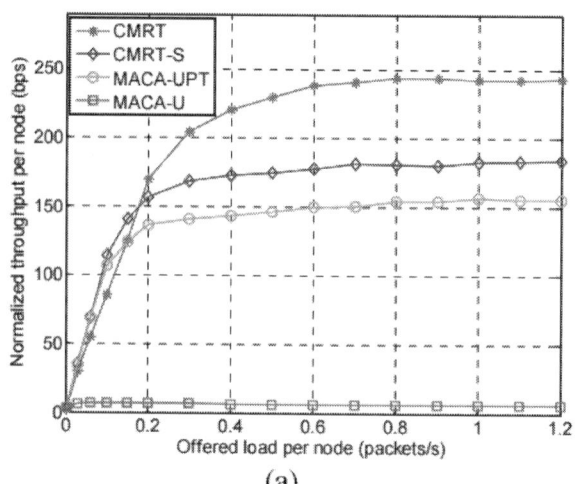

(a)

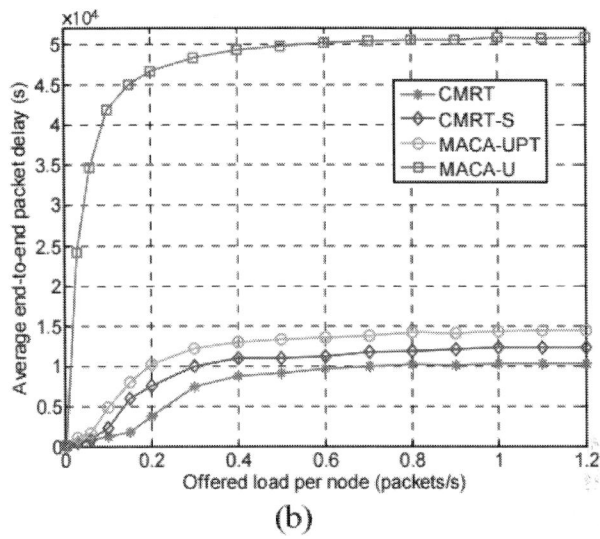

(b)

Figure 9: Performance comparisons of CMRT with other MAC protocols: (a) normalized throughput per node, and (b) average end-to-end packet delay. MACA-UPT, multiple-access collision avoidance for underwater with packet trains.

Table 3: Comparison of the features of MAC protocols

Feature \ MAC Protocol	Cascading Reservation and Transmission	Packet-Train Method	Solution for Hidden-Node Problem
CMRT	O	O	O
CMRT-S	X	O	O
MACA-UPT	X	O	X
MACA-U	X	X	X

On the other hand, when the offered load is lower than 0.2, the throughput of CMRT is slightly lower than that of CMRT-S and MACA-UPT. In CMRT, when a node is involved in multi-hop relaying as a relay node, the relay node is allowed to send only foreign data-packets and not domestic data-packets, even if it has domestic ones, in order to maintain the constant size of the data stream from source to destination. On the other hand, in hop-by-hop relaying schemes, such as CMRT-S and MACA-UPT, every hop is refreshed, so that a sender plays the role of a source all of the time and sends as many foreign and domestic data packets as possible. Thus, under a light traffic condition, CMRT does not have enough data packets to fully achieve its capability. Saturation of both the throughput and the delay with the increase in the offered load is caused by the limited buffer capacity of 300 data packets (refer to Table 2).

Figure 10 shows the system throughput *versus* the offered load, which is defined as follows:

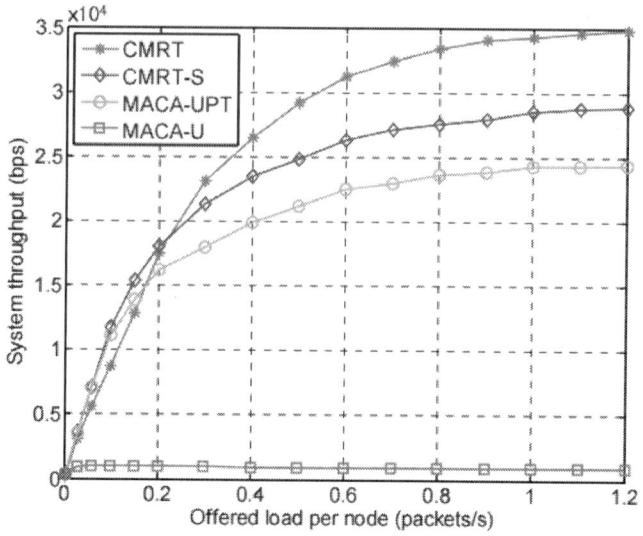

Figure 10. System throughput of CMRT in comparison with other MAC protocols.

$$S = \frac{\sum_{i=1}^{N} s_i \cdot B_{Data}}{t_{sim}}$$

(21)

where s_i is the total number of data packets successfully received by node *i*. Therefore, the system throughput includes successfully received data packets by not only final destinations, but also relay nodes. Consequently, the system throughput means the overall channel utilization by using the MAC protocol. Thus, similar to the case of normalized throughput per node, CMRT outperforms other alternatives in terms of the channel utilization.

Analysis of Hop-Delay

To provide further insight into the performance of CMRT, we analyze the delay (shown in Figure 9b) in more detail according to the number of hops between source and destination nodes that is denoted by k_{hop}. Figure 11 shows the delay of CMRT and MACA-UPT *versus* k_{hop}, when the offered load is fixed at 1.0 packets/s, which is large enough to ensure that a node always has data packets in its buffer. In our simulation model, where 36 static nodes are located in the grid of a square area, as shown in Figure 7,k_{hop} varies from one to a maximum of five. As k_{hop} increases, the difference between the delays with CMRT and MACA-UPT becomes larger, because the gain of cascading transmission is cumulative. Note that the incremental delay is not proportional to k_{hop} owing to the priority policy of oldest-first. That is, at the instant of a priority decision, the data packets with a larger k_{hop} are likely to be selected as the oldest packets that have traveled through the largest number of hops.

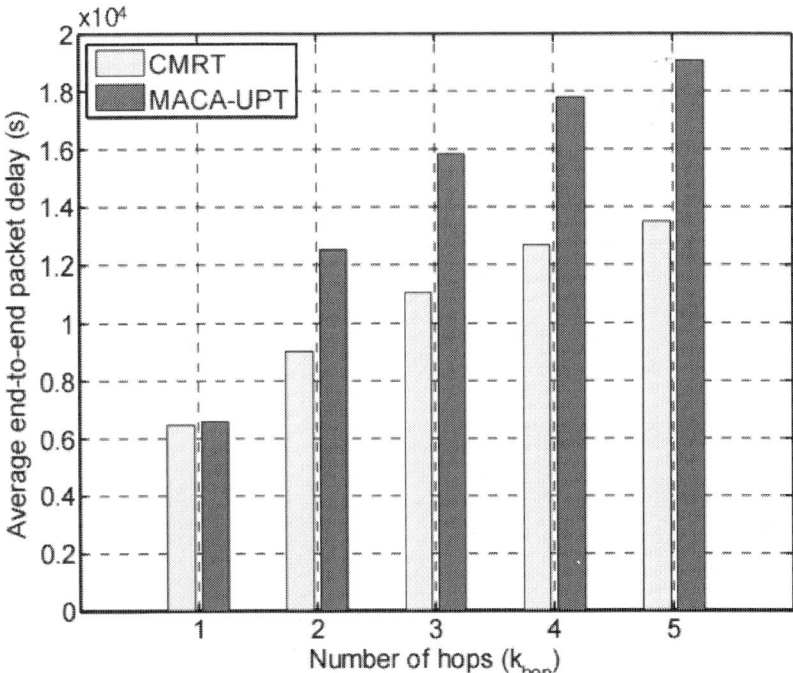

Figure 11: Delay comparison between CMRT and MACA-UPT by varying the number of hops between source and destination, k_{hop}.

Effects of Inter-Nodal Distance

Figure 12a, b shows the effects of inter-nodal distances on the throughput and the delay, under the offered load of 1.0 packets/s. As the inter-nodal distance increases, the performance of each of the protocols in terms of both throughput and delay deteriorates because the distance-related communication overhead increases accordingly. The increase in the propagation delay due to the extended distance causes an increase in the busy duration, as well as the handshaking time. Consequently, the prolonged busy duration increases the length of the Silence state, and thus, a node has less opportunity to attempt an RTS. However, we have shown that compared with MACA-UPT, CMRT could be a better solution for use in more scalable multi-hop networks.

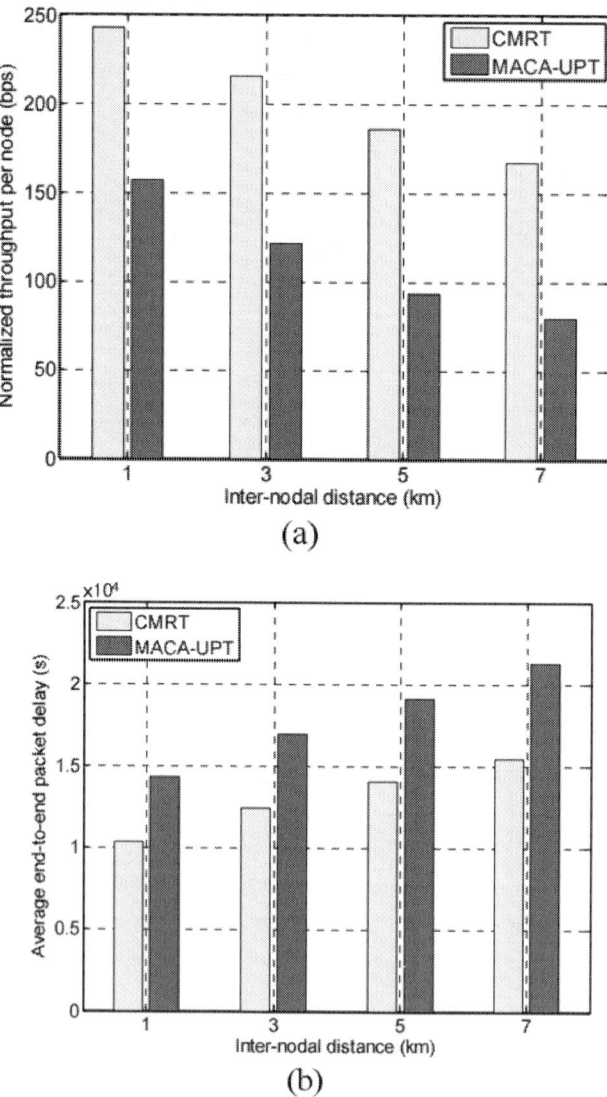

Figure 12: Effects of inter-nodal distance on CMRT and MACA-UPT: (**a**) normalized throughput per node, and (**b**) average end-to-end packet delay.

CONCLUSIONS

This paper has discussed the challenges posed by the long propagation delay in the underwater acoustic channel that need to be considered when designing an underwater channel MAC protocol and achieving the benefits of multi-hop relay. On the basis of these considerations, a cascading multi-hop reservation and transmission MAC protocol named CMRT has been proposed. To reduce the time-related overhead caused by the propagation delay, CMRT makes a relay node start handshaking for the next hop, while the handshaking for the previous hop is in progress. With this concurrent relaying, the flow of control and data packets starting from a source cascades to the destination without stopping at any relay nodes. In addition, CMRT is able to reduce the control packet exchange time by utilizing the backward overhearing of a control packet forwarded by the next hop relay node as a response. To prevent unexpected collisions caused by the hidden-node problem, CMRT postpones data transmission until potential interferers enter the silent mode by recognizing the channel to be reserved. Furthermore, to improve channel utilization, CMRT adopts a packet-train method. Computer simulation shows that CMRT outperforms other well-known underwater MAC protocols, such as MACA-U and MUAC-UPT in terms of both throughput and delay. In further works, the following problems will be investigated: (1) a time-efficient acknowledgment scheme; (2) improvement of spatial fairness between sensor nodes; and (3) protocol evaluation under more realistic underwater acoustic channel conditions.

This work was supported in part by the Defense Acquisition Program Administration and Agency for Defense Development under the contract, UD130007DD, and in part by Basic Science Research Program through the National Research Foundation of Korea funded by the Ministry of Education, Science and Technology (NRF-2012R1A1A4A01).

AUTHOR CONTRIBUTIONS

Jae-Won Lee developed the CMRT MAC protocol, run computer simulations, and wrote the manuscript. Ho-Shin Cho supervised the works and participated in data analysis and revision process.

REFERENCES

1. Zhang, B.; Sukhatme, G.S.; Requicha, A.G. Adaptive Sampling for Marine Microorganism Monitoring. Proceedings of the IEEE/RSJ International Conference on Intelligent Robots and System (IROS 2004), Sendai, Japan, 28 September–2 October 2004; pp. 1115–1122.

2. Akyildiz, I.F.; Pompili, D.; Melodia, T. Underwater Acoustic Sensor Networks: Research Challenges. *Ad Hoc Netw.* **2005**, *3*, 257–279.

3. Vieira, L.F.M.; Kong, J.; Lee, U.; Gerla, M. Analysis of Aloha Protocols for Underwater Acoustic Sensor Networks. Proceedings of the First ACM International Workshop on Underwater Networks (WUWNet 2006), Los Angeles, CA, USA, 25 September 2006.

4. Syed, A.; Wei, Y.; Heidemann, J.; Krishnamachari, B. Understanding Spatio-Temporal Uncertainty in Medium Access with ALOHA Protocols. Proceedings of the Second Workshop on Underwater Networks, Montreal, QC, Canada, 14 September 2007; pp. 41–48.

5. Chirdchoo, N.; Soh, W.S.; Chua, K.C. Aloha-based MAC Protocols with Collision Avoidance for Underwater Acoustic Networks. Proceedings of the 26th Annual IEEE Conference on Computer Communications (INFOCOM 2007), Anchorage, AK, USA, 6–12 May 2007; pp. 2271–2275.

6. Kleinrock, L.; Tobagi, F. Packet Switching in Radio Channels: Part 1—Carrier Sense Multiple-Access Modes and their Throughput-Delay Characteristics. *IEEE Trans. Commun.* **1975**, *23*, 1400–1416.

7. Karn, P. MACA: A New Channel Access Method for Packet Radio. Proceedings of the 9th Computer Networking Conference, London, Ontario, Canada, 22 September 1990; pp. 134–140.

8. Molins, M.; Stojanovic, M. Slotted FAMA: A MAC Protocol for Underwater Acoustic Networks. Proceedings of the IEEE OCEANS'06 Asia Pacific, Singapore, 16–19 May 2006; pp. 16–19.

9. Peleato, B.; Stojanovic, M. Distance Aware Collision Avoidance Protocol for Ad-hoc Underwater Acoustic Sensor Networks. *IEEE Commun. Lett.* **2007**, *11*, 1025–1027.

10. Petrioli, C.; Petroccia, R.; Stojanovic, M. A Comparative Performance Evaluation of MAC Protocols for Underwater Sensor Networks. Proceedings of the MTS/IEEE OCEANS 2008, Kobe, Japan, 8–11 April 2008; pp. 1–10.

11. Syed, A.; Ye, W.; Heidemann, J. Comparison and Evaluation of the T-Lohi MAC for Underwater Acoustic Sensor Networks. *IEEE J. Sel. Areas Commun.* **2008**, *26*, 1731–1743.

12. Casari, P.; Tomasi, B.; Zorzi, M. A Comparison between the Tone-lohi and Slotted FAMA MAC Protocols for Underwater Networks. Proceedings of the MTS/IEEE OCEANS 2008, Quebec City, Canada, 15–18 September 2008; pp. 1–8.

13. Ng, H.H.; Soh, W.S.; Montani, M. MACA-U: A Media Access Protocol for Underwater Acoustic Networks. Proceedings of the IEEE Global Telecommunications Conference 2008, New Orleans, LO, USA, 30 November–4 December 2008.

14. Liao, W.H.; Huang, C.C. SF-MAC: A Spatially Fair MAC Protocol for Underwater Acoustic Sensor Networks. *IEEE Sens. J.* **2012**, *12*, 1686–1694.

15. Guo, X.; Frater, M.R.; Ryan, M.J. Design of a Propagation-Delay-Tolerant MAC Protocol for Underwater Acoustic Sensor Networks. *IEEE J. Ocean. Eng.* **2009**, *34*, 170–180.

16. Noh, Y.; Lee, U.; Han, S.; Wang, P.; Torres, D.; Kim, J.; Gerla, M. DOTS: A Propagation Delay-Aware Opportunistic MAC Protocol for Mobile Underwater Networks. *IEEE Trans. Mob. Comput.* **2014**, *13*, 766–782.

17. Chirdchoo, N.; Soh, W.S.; Chua, K.C. MACA-MN: A MACA-based MAC Protocol for Underwater Acoustic Networks with Packet Train for Multiple Neighbors. Proceedings of the IEEE Vehicular Technology Conference, Singapore, 11–14 May 2008; pp. 46–50.

18. Ng, H.H.; Soh, W.S.; Montani, M. BiC-MAC: Bidirectional-Concurrent MAC Protocol with Packet Bursting for Underwater Acoustic Networks. Proceedings of the MTS/IEEE OCEANS 2010, Seattle, WA, USA, 20–23 September 2010.

19. Ng, H.H.; Soh, W.S.; Montani, M. An Underwater Acoustic MAC Protocol using Reverse Opportunistic Packet Appending. *Comput. Netw.* **2013**, *57*, 2733–2751.

20. Chirdchoo, N.; Soh, W.S.; Chua, K. RIPT: A Receiver-Initiated Reservation-based Protocol for Underwater Acoustic Networks. *IEEE J. Sel. Areas Commun.* **2008**, *26*, 1744–1753.

21. Xie, P.; Cui, J.H. R-MAC: An Energy-efficient MAC Protocol for Underwater Sensor Networks. 187–198.

22. *LAN/MAN (IEEE 802) Std. Comm. Wireless LAN Media Access Control (MAC) and Physical Layer (PHY) Specifications*; IEEE: New York, NY, USA; September; 1999.

23. LinkQuest Inc. Available online: http://www.link-quest.com (accessed on 30 August 2014).

CITATION

Jae-Won Lee and Ho-Shin Cho, Cascading Multi-Hop Reservation and Transmission in Underwater Acoustic Sensor Networks, doi: 10.3390/ s141018390.

Index